T0393841

Introduction to Aviation Operations Management

Airline operations are large, complex, and expensive. *Introduction to Aviation Operations Management* attempts to systematically present the overall scenario of aviation industry and airline practices. Furthermore, concepts, strategies, and issues prevailing in the aviation industry are addressed through numerous operations management and optimization approaches. The book aims to provide readers with an insight into aviation industry practices with respect to airport management, resource allocation, airline scheduling, disruption management, and sustainability which are significant for day-to-day aviation operations.

Features:

- Presents operations management perspectives in the aviation sector.
- Discusses global scenarios of aviation industry and airline practices.
- Concepts are explained through operations management and optimization approaches.
- Discusses airport management, resource allocation, airline scheduling, and disruption management issues.
- Includes standard practices and issues related to the aviation industry.

This book is aimed at senior undergraduate students pursuing programs related to the aviation industry and operations management.

Introduction to Aviation Operations Management

Sheikh Imran Ishrat
Zahid Akhtar Khan
Arshad Noor Siddiquee

CRC Press is an imprint of the
Taylor & Francis Group, an **informa** business

Cover image: © Shutterstock

First edition published 2023
by CRC Press
6000 Broken Sound Parkway NW, Suite 300, Boca Raton, FL 33487-2742

and by CRC Press
4 Park Square, Milton Park, Abingdon, Oxon, OX14 4RN

CRC Press is an imprint of Taylor & Francis Group, LLC

ISBN: 9781138303218 (hbk)
ISBN: 9781032457000 (pbk)
ISBN: 9780203731338 (ebk)

DOI: 10.1201/9780203731338

Typeset in Times
by Newgen Publishing UK

Dedication

This book is dedicated to the contribution of women in the advancement of the aviation industry and to the spirit of Amelia Earhart for inspiring generations of women in aviation practice and research.

Contents

List of Abbreviations xiii
Preface xv
About the Authors xvii

Chapter 1 History of Civil Aviation 1

1.1 Introduction 1
1.2 Before 1900: Early Attempts 1
1.3 1900–1950: The Evolution 3
1.4 1950–2000: Connecting the World 5
1.5 2000 and Beyond: Changing Landscapes 6
1.6 Conclusion 7
Chapter Questions 8
References 8

Chapter 2 Global Aviation Operations 9

2.1 Introduction 9
2.2 Aviation Regulatory Bodies 10
2.2.1 International Civil Aviation Organization 10
2.2.2 International Air Transport Association 11
2.2.2.1 Western Hemisphere (WH) 12
2.2.2.2 Eastern Hemisphere (EH) 12
2.2.2.3 Pacific (PA) 12
2.2.2.4 Atlantic (AT) 13
2.2.2.5 Atlantic/Pacific (AP) 13
2.2.2.6 South Asia (SA) 13
2.2.2.7 Far East (FE) 13
2.2.2.8 Russia (RU) 14
2.2.2.9 Trans-Siberian (TS) 14
2.2.3 European Aviation Space Agency 14
2.2.4 Federal Aviation Administration 14
2.3 Airport Locations and Categories 15
2.3.1 Western Hemisphere (TCA 1) 15
2.3.1.1 North and Central America 15
2.3.1.2 South America 16
2.3.2 Eastern Hemisphere (TCA 2) 18
2.3.2.1 Europe 18
2.3.2.2 Africa and the Middle East 19
2.3.3 Eastern Hemisphere (TCA 3) 21
2.3.3.1 South Asian Subcontinent 21

2.3.3.2 Southeast Asia, China, Japan, and Korea (TCA 3) 23
2.3.3.3 Southwest Pacific (TCA 3) 23
2.4 Airlines 25
2.4.1 Hub-and-Spoke Network 25
2.4.2 Point-to-Point Network 26
2.5 Consumers 27
2.6 Impact of COVID-19 on the Aviation Industry 28
2.7 Conclusion 29
Chapter Questions 30
References 30

Chapter 3 Civil Aviation Landscape in India 33

3.1 Introduction 33
3.2 Civil Aviation Regulatory Bodies 34
3.2.1 Continuing Airworthiness 34
3.2.2 Aircraft Certification 35
3.2.3 Air Navigation Services 35
3.2.4 Aircraft Operations 35
3.2.5 Legal Affairs 35
3.2.6 Personal Licensing 35
3.2.7 Aerodromes and Ground Aids 36
3.2.8 Administration 36
3.2.9 Flying Training and Sports 36
3.2.10 Air Operator Certification and Management 36
3.2.11 Investigation and Prevention 36
3.2.12 Information Technology 36
3.2.13 Training 37
3.2.14 International Cooperation 37
3.2.15 Surveillance and Enforcement 37
3.2.16 State Safety Program 37
3.3 Airports in India 37
3.3.1 Domestic Airports 38
3.3.2 International Airports 38
3.4 Airline Operators in India 44
3.4.1 The Early Days 44
3.4.2 Major Developments 44
3.4.3 Emergence of Low-Cost Carriers (LCC) 46
3.5 Conclusion 48
Chapter Questions 48
References 49

Chapter 4 Aviation Supply Chains 51

4.1 Introduction 51
4.2 Supply Chain Management 51
4.2.1 Strategies and Challenges 53
4.2.2 Procurement 54
4.2.3 Manufacturing 56
4.2.4 Distribution 60
4.3 Aviation Supply Chains 61
4.3.1 Airline Supply Chain 61
4.3.1.1 Equipment Manufacturers 62
4.3.1.2 Fuel Supplies 63
4.3.1.3 Catering Services 63
4.3.1.4 Agents 64
4.3.2 Airport Supply Chain 66
4.4 Conclusion 68
Chapter Questions 68
References 68

Chapter 5 Airport Operations 71

5.1 Introduction 71
5.2 Operational Aspects 72
5.2.1 Airside Operations 72
5.2.1.1 Apron 73
5.2.1.2 Taxiway 73
5.2.1.3 Runway 74
5.2.1.4 ATC Operations 74
5.2.2 Landside Operations 75
5.3 Airport Competition 76
5.3.1 Airport Monopoly 76
5.3.2 Airport Capacity 76
5.3.3 Airport Location 77
5.3.4 Airport Service Quality 78
5.3.5 Passenger Considerations 79
5.4 Conclusion 80
Chapter Questions 80
References 80

Chapter 6 Airline Operations 83

6.1 Introduction 83
6.2 Airline Industry 84
6.2.1 Airline Schedule Construction 84
6.2.2 Flight Schedule 85
6.2.3 Fleet Assignment 85

6.2.4 Aircraft Routing 86
6.2.5 Crew Scheduling 86
6.3 India-Based Domestic Network 88
6.3.1 Flight Network 88
6.3.2 Aircraft Routings 89
6.3.3 Crew Pairings 92
6.4 Other Scheduling Techniques 97
6.5 Conclusion 98
Chapter Questions 98
References 98

Chapter 7 Airline Disruption Management – I 101

7.1 Introduction 101
7.2 Disruption Management 101
7.3 Types of Disruption 102
7.4 Schedule Recovery from Disruptions 104
7.5 Aircraft Recovery Approaches 105
7.5.1 Solution Approaches Based on Network Flows 105
7.5.2 Solution Approaches Based on Time–Space Networks 106
7.5.3 Solution Approaches Based on Time–Band Networks 109
7.5.4 Set Partitioning Models Formulated on Connection Networks 110
7.5.5 Heuristic Approaches 111
7.5.6 Other Approaches 113
7.6 Crew Recovery Approaches 114
7.6.1 Crew Recovery with Fixed Fight Schedule 114
7.6.2 Crew Recovery with Flight Cancellations 116
7.6.3 Crew Recovery with Departure Delays 117
7.7 Passenger and Integrated Recovery 118
7.8 Conclusion 119
Chapter Questions 120
References 120

Chapter 8 Airline Disruption Management – II 123

8.1 Introduction 123
8.2 The Concept of Disruption Neighborhood 123
8.3 Initial Neighborhood 125
8.3.1 Resources of the Disrupted Flight: Aircraft 127
8.3.1.1 Entry Time and Entry Port 128
8.3.1.2 Exit Time and Exit Port 128

8.3.2 Resources of the Disrupted Flight: Crew 128
8.3.2.1 Entry Time and Entry Port 128
8.3.2.2 Exit Time and Exit Port............................ 128
8.3.3 Resources of the Affected Flights 128
8.4 Expansion of Neighborhood 130
8.5 Network Creation and Column Generation 133
8.5.1 Aircraft Network 135
8.5.1.1 Aircraft Routing 135
8.5.1.2 Aircraft Columns...................................... 136
8.5.2 Captain Network.. 137
8.2.2.1 Captain Pairings 137
8.2.2.2 Captain Columns....................................... 138
8.5.3 First Officer Network 139
8.5.3.1 First Officer Pairings 140
8.5.3.2 First Officer Columns............................... 142
8.6 Set Partitioning Formulation.. 143
8.7 Rolling Time Horizon Recovery 143
8.8 Other Recovery Considerations 145
8.9 Conclusion 148
Chapter Questions 148
References 151

Chapter 9 Sustainability in the Aviation Industry 153

9.1 Introduction.. 153
9.2 Environmental Sustainability 154
9.2.1 Air Pollution 154
9.2.2 Noise Pollution... 155
9.2.3 Aviation Fuel 156
9.2.4 Water Pollution... 158
9.2.5 Climate Change 158
9.3 Social and Economic Sustainability in Aviation 159
9.4 Sustainability in Airport Operations 160
9.5 Sustainability Reporting in Aviation Industry........................ 161
9.6 Challenges in Aviation Sustainability 162
9.7 Achieving Sustainability in Aviation Industry 163
9.7.1 Technological Improvements 163
9.7.2 Fuel Options 164
9.7.3 Aviation Policy 164
9.7.4 Waste Management Aspects.................................... 165
9.8 Conclusion 166
Chapter Questions 166
References 166

Chapter 10 Comparison of Airline and Railway Operations 173

10.1 Introduction 173
10.2 Scheduling in Railway Industry 173
10.2.1 Sequential Scheduling 174
10.2.2 Robust Scheduling 175
10.3 Disruption Management in Railways 175
10.3.1 Timetable Adjustments 177
10.3.2 Rolling Stock Re-scheduling 178
10.3.3 Crew Re-scheduling 179
10.4 Airline and Railway Scheduling: A Comparison 179
10.4.1 Similarities 180
10.4.1.1 Schedule Construction 180
10.4.1.2 Recovery Procedure 180
10.4.1.3 Solution Approaches 180
10.4.2 Differences 181
10.4.2.1 Operational 181
10.4.2.2 Technical 181
10.4.2.3 Recovery Complexity 182
10.4.2.4 Passenger Issues 182
10.5 Conclusion 182
Chapter Questions 183
References 183

Appendices 185

Index 191

Abbreviations

AAI	Airports Authority of India
ACI	Airport Council International
AED	Aero Engineering Division
AEU	Aviation Environment Unit
ALD	Aero Laboratories Division
AME	Aircraft Maintenance Engineers
AOCMB	Air Operator Certification and Management Bureau
AT	Atlantic
ATD	Air Transport Division
ATC	Air Traffic Control
AP	Atlantic Pacific
BOAC	British Overseas Airways Corporation
COVID-19	Coronavirus Disease
CPCB	Central Pollution Control Board
DGCA	Directorate General of Civil Aviation
EASA	European Aviation Space Agency
EH	Eastern Hemisphere
FAA	Federal Aviation Administration
FE	Far East
FSC	Full-Service Carrier
GDP	Gross Domestic Product
HAL	Hindustan Aeronautics Limited
IAAI	International Airports Authority of India
IATA	International Air Transport Association
IBEF	India Brand Equity Foundation
ICAO	International Civil Aviation Organization
INR	Indian Rupee
KLM	Koninklijke Luchtvaart Maatschappij (Dutch Airlines)
LCC	Low-Cost Carrier
MERS	Middle East Respiratory Syndrome
MCA	Ministry of Civil Aviation
NAA	National Airports Authority
NASA	National Aviation Space Agency
PA	Pacific
PCR	Polymerase Chain Reaction
SA	South Asia
SARP	Standards and Recommended Practices
SARS	Severe Acute Respiratory Syndrome

TCA Traffic Conference Area
TS Trans-Siberian
WE Western Hemisphere
WWI World War I
WWII World War II

Preface

First and foremost, no amount of gratitude is enough to thank God (The Almighty) for providing the opportunity to work on this book.

The aspiration to fly has existed within the human beings for centuries. From mimicking birds in flight to developing sophisticated technologies which finally made aviation possible, it has been a long and impactful journey that has changed the course of human history. It is the relentless efforts, individual and collective, that have made possible not only the operation of complex flying machines but built an intricate infrastructure of communication and cooperation spanning the entire globe.

While there is a plethora of textbooks, reference books, and other reading material about the aviation industry; capturing mathematical aspects, business objectives, sustainability considerations, and industry regulations, or even a combination of these, there is a dearth of material covering operations specifically within the context of Indian aviation. This book aims to provide a holistic view of the Indian aviation industry through the lens of operations management. The text is primarily geared for senior undergraduate students as well as graduate students pursuing business and engineering programs. For researchers and practitioners, the book may serve as a primitive reference to get an overarching view of the aviation industry. This book is a result of an in-depth study conducted on airline disruption management, the courses taught by the authors across different programs of study and other sources of information regarding the aviation industry. This subsequently evolved to encompass the aviation industry operations. The objective of this text is twofold. First, it covers the vast landscape of the industry and attempts to explore its different aspects at play. Second, it presents the application of quantitative techniques and qualitative approaches within the context of aviation operations in India.

The authors have considerable experience in teaching undergraduate and graduate level courses in the areas of operations and supply chain management as well as in industrial and mechanical engineering. Furthermore, based on extensive research of the extant literature surrounding aviation operations and industry practices, content is sourced and duly cited within the text. The topics included in the book are structured with care and can be covered in the presented sequence. To that end, a balanced approach was adopted to ensure the coverage of a range of aviation operations in a lucid yet insightful manner within the text. Moreover, many technical concepts and details have been explained with the aid of illustrations which is a distinctive feature of this book. Finally, discussion questions are provided at the end of each chapter for a better understanding of the content and to underscore the key takeaways from the topics covered.

Sincere gratitude is extended to Prof. habil. Matthias Ehrgott, Emeritus Prof. David Ryan, and Paul Keating for sharing their expertise and insights on airline scheduling and recovery operations. Their guidance and support are much appreciated. The researchers whose research contributions have helped shape this book are also thanked. Furthermore, FlightConnections.com and AirportCodes.com are acknowledged for giving the permission to use the figures and data from their websites which complemented well with the content of the book. Last but not the least, Tipu Sultan deserves a special mention for the painstaking effort to read through the manuscript draft and suggesting improvements for future editions.

Finally, the authors are indebted to their families for the motivation they provided in completing this book.

Sheikh I. Ishrat, *Christchurch*
Zahid A. Khan, *New Delhi*
Arshad N. Siddiquee, *New Delhi*

About the Authors

Sheikh Imran Ishrat, has a teaching and research career spanning more than a decade. He has been teaching Operations Management, Supply Chain Management, and Industrial Engineering courses across Business and Engineering programs in New Zealand and Asia. Dr. Ishrat received his PhD in Engineering from Massey University (New Zealand), MEng in Engineering Management from University of Ottawa (Canada) and an MSc in Operations Research from Aligarh Muslim University (India). He has been supervising research projects, published articles, and presented his work in reputed international conferences including IFORS, IISE, IEEE, and APQO. Dr. Ishrat's research was placed in the Massey University's prestigious Dean's List of Exceptional Doctoral Theses and received the Excellent Paper Presentation award in ICITM at University of Cambridge (UK).

Zahid Akhtar Khan is Professor in the Department of Mechanical Engineering at Jamia Millia Islamia (A Central University), New Delhi, India. He has more than 30 years of teaching and research experience. He received his PhD in Mechanical Engineering, specialization in Industrial Engineering, Jamia Millia Islamia, New Delhi, India. He has been working in the area of optimization of manufacturing and service processes. He has supervised ten Doctoral Theses and many M. Tech. Dissertations so far and currently he is supervising several Doctoral Theses. He has published more than 200 research papers in reputed International/National journals and conferences so far. His current Google Scholar Citations is 4,042, *h*-index is 32and *i*-10 index is 77. He has also co-authored six books related to engineering discipline and two monographs as well.

Arshad Noor Siddiquee is Professor in the Department of Mechanical Engineering at Jamia Millia Islamia (A Central University), New Delhi, India. He has more than 25 years of teaching and research experience. He received his PhD in Production Engineering from IIT Delhi, India. His major research interests include aerospace materials and production. He has supervised eight Doctoral Theses and many M. Tech. Dissertations so far and currently he is supervising several Doctoral Theses. He has published more than 250 research papers in reputed International/National journals and conferences so far. His current Google Scholar Citations is 3,935, *h*-index is 34and *i*-10 index is 91. He has also co-authored six books related to engineering discipline and two monographs as well.

1 History of Civil Aviation

CHAPTER OBJECTIVES

At the end of this chapter, you will be able to

- Know the different stages of development in the civil aviation industry.
- Understand the advancements in the civil aviation operations.
- Know the major events in the history of civil aviation.
- Know about the significant individuals and their contributions.
- Get an overview of the timeline of the developments in the civil aviation industry.

1.1 INTRODUCTION

The history of aviation has existed from time immemorial. In the last couple of centuries, recorded evidence suggests that humanity's trysts with flight have spanned a range of methodologies. Early flying instances indicate humans trying to imitate birds by jumping off elevated places and flapping artificial wings attached to their bodies. However, since the beginning of the 20th century, the efforts to fly were based on scientific approaches rather than merely replicating the actions of the Aves. This led to numerous possibilities for exploring the unknown. A significant development in the trajectory of the aviation industry is observed after the first successful flight of a controlled, powered machine – the aircraft. Decades later, the modern-day aviation industry can be classified into different categories based on their operations such as civil aviation (commercial airlines), military operations (combat fighters), and general (personal flying) aviation. However, in this chapter, the focus is on the civil aviation aspects of the aviation industry.

1.2 BEFORE 1900: EARLY ATTEMPTS

Humanity's fixation for flying has existed for a long time. Anecdotes about the quest to conquer the skies and reach out to foreign and faraway lands are prevalent in different mythos and formed a common folklore across cultures. Early civilizations such as the Chinese used to fly kites, depicting the human aspiration to fly. Similarly,

DOI: 10.1201/9780203731338-1

many attempts have been made in different eras and different areas to achieve the then-impossible task of flight.

In this pursuit, one of the early recorded events includes tower jumps made by individuals, such as Abbas Ibn Firnas in 875 who built a mechanism to glide from the tower top in Andalusia before descending to the ground after a rough flight (Altuntas et al., 2019; Grant, 2017, p. 10). Later, numerous attempts to fly were made using hot air balloons in which the efforts of Leonardo da Vinci (1452–1519) and the Montgolfier brothers (1740–1810) are considered significant in the shaping of future flights. Da Vinci also developed a flapping wings mechanism imitating the movements of a bird. However, in the late 18th century advancements were made by the Montgolfier brothers (Joseph-Michel and Jacques-Étienne) regarding lighter-than-air flight displays in 1783 using the hot air balloons (Grant, 2017, p. 11). The first flight carrying a human being was also recorded in 1783 when Jean-François Pilâtre de Rozier (1754–1785) and Francois d'Arlandes (1742–1809) covered eight kilometers on a balloon devised by the Montgolfier brothers (Altuntas et al., 2019). By the 19th century, hot air balloons became quite popular, and in 1852 Henri Giffard (1825–1882) operated the first controlled, powered balloon in France (Grant, 2017, p. 13).

Similarly, efforts were made to power heavier-than-air flights and by the mid-19th century experiments using steam-powered flights gained traction. George Cayley (1773–1857) operated the first manned heavier-than-air flight using a glider in 1853. By the end of the 19th century Otto Lilienthal (1848–1896) successfully glided a controlled flight on a heavier-than-air machine demonstrating the advances made in the early phases of aviation (Anderson, Jr., 2002 as cited in Altuntas et al., 2019). Table 1.1 presents major events in the early days of aviation operations where individual attempts were made using different approaches to fly.

By the end of the 19th century, there was a realization among aviators to focus on and develop heavier-than-air controlled flying mechanisms to overcome the shortcomings of lighter-than-air flights. Therefore, in the beginning of the 20th century, significant advances were made in the development of controlled and powered airplanes (Appendix C) which were based on scientific approaches as presented in Section 1.3.

TABLE 1.1
Significant events and major contributors in early aviation operations

Timeline	Name	Major Events
400 B.C.		The Chinese flew the kites.
875	Abbas Ibn Firnas	Made the first flight using a flying mechanism.
1485	Leonardo da Vinci	Designed a wing-flapping machine.
1783	Montgolfier brothers	Launched the first hot air balloon.
1783	de Rozier & d'Arlandes	First balloon flight carrying humans.
1849	George Cayley	First manned heavier-than-air flight in a glider.
1852	Henri Giffard	Operated first controlled, powered balloon flight.
1891	Otto Lilienthal	Used the glider for controlled heavier-than-air flight.

1.3 1900–1950: THE EVOLUTION

Within the aviation industry, the first half of the 20th century has observed more significant developmental changes than hardly any other span of time. By the first decade, numerous attempts and claims were made to fly a powered machine. Among others, the most prominent instance was in 1901–1902, when an unconfirmed account of flying an airplane by Gustave Whitehead (1874–1927) was reported (Eves, 2018). Also, Sam Langley (1834–1906) piloted a powered machine in 1903 but was unsuccessful in his attempt. However, the breakthrough to break through the air by defying gravity for about a minute and flying for more than 250 meters came in 1903 when Wilbur Wright (1867–1912) and Orville Wright (1871–1948) piloted the first controlled, powered flight in the history of mankind (Grant, 2017, p. 20). One of the major reasons for the success of the Wright brothers can be attributed to their novel approach in maintaining the required balance and control of the airplane through the air by combining lightness with strength, similar to pedalling a bicycle (Grant, 2017, p. 23). By 1905, the Wright brothers had conducted more than 80 public flights – the longest lasting over 30 min – demonstrating their flying skills, control, and superior airplane design (Altuntas et al., 2019; Grant, 2017, p. 28). In 1906, Alberto Santos-Dumont (1873–1932) also successfully made the first powered flight by covering more than 200 meters in a heavier-than-air machine and pioneered aviation in Europe (Grant, 2017, p. 31). Later, in 1908, the Wright brothers further demonstrated a sustained and controlled display of powered machine by flying it 110 meters above the ground for over 2 h in France (Grant, 2017, p. 33). With the combined efforts of early generation aviators, engineers, entrepreneurs, and scientists, suitable changes were incorporated in the design of powered machines leading to considerable advancements in the aviation industry. As a result, by the end of the first decade of the 20th century, Louis Blériot (1872–1936) flew over the English Channel in 1909 in over 30 min, covering around 35 kms at an average speed of 65 km/h (Grant, 2017, p. 40). The beginning of the second decade of the 20th century ushered in a number of milestones, including the speed record (Hélène Dutrieu in 1911), first seaplane take-off (Curtiss-Ellyson in 1911), first mail service (Allahabad – India in 1911) and the first parachute jump (Albert Berry in 1912) which were heralded as major events that revolutionized the aviation industry and paved the way forward for the next generation aviators. By 1915, the aviation industry was evolving fast in Europe, and countries such as Britain, France, and Germany had made considerable progress in establishing their aviation industry. Also, long-distance flights were operational too, as evident from Sikorsky's 2,600 km round-trip flight in Russia (Grant, 2017, p. 63). Apart from considering flying for civilian use, another application of the airplane was observed for military purposes during the first Balkan War (1912–1913) and World War I (1914–1918), adding a new dimension in the warfare (Altuntas et al., 2019). WWI became the catalyst for the mass production of aircraft and demonstrated its significance. As a result, aircraft of various capabilities and designs came into prominence and were used frequently for combat operations demonstrating their ability, reliability, and established their utility in the aviation industry. After WWI, long-range flights were operational as recorded in two instances of transatlantic flights. In the first, John Alcock (1892–1919) and Arther Brown (1886–1948) crossed the North

Atlantic Ocean from the shores of Canada and reached Ireland in 1919 (Grant, 2017, p. 111). Whereas in the second instance, Carlos Coutinho (1869–1959) and Artur Cabral (1881–1924) flew the other way around over South Atlantic Ocean in 1922 from Portugal to Brazil. Another major instance of crossing the Atlantic Ocean was reported when Charles Lindbergh (1902–1974) flew solo in 1927 on a non-stop flight from New York to Paris. Hélène Dutrieu (1877–1961), a Belgian who is one of the early females to receive a pilot' license became the first female in the aviation history to fly along a passenger in 1910 and two years later, in 1912, Harriet Quimby (1875–1912) who was the first American female to get a pilot's license, became the first woman to cross the English Channel (Grant, 2017, p. 52). The success of early generation woman aviators inspired many other women to take up on flying. For instance, in 1921, Bassie Coleman (1892–1926) gained an international pilot's license and Amelia Earhart (1897–1937) became the first female pilot to cross the Atlantic Ocean solo in 1932. Before her disappearance in 1937, Earhart attempted to fly across the globe and in the process captured worldwide attention on the vast range of untapped possibilities surrounding aviation.

Even in the earlier days of aviation, people with profound insight realized the potential of the airplane which was not confined to just flying alone but also introduced a new dimension – the use of airplanes for commercial purposes across national boundaries. Therefore, in 1910, a conference, attended by 18 European States, was called in Paris to discuss an international air law code to lay the basic aviation governing principles (ICAO, n.d.). As a result, by 1920 civil aviation enterprises were established in many European countries which were providing air transport services. For instance, German airline Deutsche Luft Reederei (became Lufthansa in 1934) operated the first passenger service in 1919 and in the same year, the Dutch airline (KLM) was formed (Grant, 2017, p. 134). Also, the first daily non-stop air service between London and Paris started in 1919. Later, Air France (in 1933) and British Overseas Airways Corporation was found in 1939. In 1938, Lufthansa carried out the first commercial transatlantic flight and carried 26 passengers from Berlin to New York (Grant, 2017, p. 149). During the same period, in the United States the aviation operations progressed considerably. Busy airports such as Chicago and Newark were operating more than 50 take-offs and landings every hour, and by the end of the third decade, American airlines were carrying three million passengers annually (Grant, 2017, p. 142).

To cover large distances in less time, the need for speed, technical, and operational reliability was the catalyst for improved aircraft designs which resulted in pressurized cabins for high-altitude flying. To navigate the flights, radio beams and instrument flying were used by the pilots, and by the mid-1930s commercial aircraft were equipped with the ground air traffic controllers for communication purposes. In-flight services were also introduced to enhance passenger flying experience and, despite certain limitations, the notion that 'the aircraft is the most constrained form of mass transport since the slave ships' (Grant, 2017, p. 397) became a thing of the past.

The Second World War (1939–1945) established the significance of the aircraft in many aspects. Post WWII, among other facets, commercial aviation and transportation of goods were the major operations that benefitted most in terms of technological

TABLE 1.2
Major events and contributors in the aviation operations between 1900 and 1950

Timeline	Name	Major Events
1903	Wright Brothers	Flew first controlled powered flight in history.
1906	Alberto Dumont	Operated first heavier-than-air flight in Europe.
1909	Louis Blériot	Flew across the English Channel.
1910	Hélène Dutrieu	First women to fly with a passenger.
1911	Curtiss-Ellyson	First seaplane take-off.
1911	Allahabad, India	First air mail service.
1912	Albert Berry	First parachute jump.
1913	Balkan War	Use of aircraft for combat purposes.
1919	Alcock & Brown	Flew non-stop across the Atlantic Ocean.
1919	Lufthansa	First commercial passenger service.
1927	Charles Lindbergh	Solo non-stop flight between New York and Paris.
1932	Amelia Earhart	First women to fly on a solo transatlantic flight.
1938	Lufthansa	First commercial transatlantic flight (Berlin to New York).
1947	Chuck Yeager	Sound barrier was broken.

advances and scale of aviation operations. By the end of the 1940s, the aviation industry was developed to the extent that the sound barrier, i.e., aircraft reaching the speed of the sound (1230 km/h), also known as Mach 1, was broken by Chuck Yeager in California, United States (Grant, 2017, p. 266). This was a significant and historic achievement in the aviation operations depicting the exponential progress made in four decades since the first manned flight operated by the Wright brothers. Table 1.2 presents major events during the period between 1900 and 1950 which evolved the aviation industry and established global aviation operations.

The first half of the 20th century is considered as the golden age of aviation by many accounts. The achievements and developments attained during this period were critical in establishing the future civil and military aviation operations.

1.4 1950–2000: CONNECTING THE WORLD

By the 1950s, airline operators amassed considerable knowledge and experience of aviation industry operations. Furthermore, technological advancements coupled with innovation and research and infrastructural development helped the industry grow exponentially. As a result, far-flung places were reached which otherwise would have been difficult to access via conventional transportation. Also, institutional support from government bodies and other industry stakeholders helped commercial aviation operators to develop their network beyond regional or national boundaries. As a result, many airlines based in Europe and other parts of the world developed their international flight networks saving them high logistics and infrastructural expenses. Especially, with the development of aviation infrastructure across both sides of the Atlantic, by the end of the 1950s more people made the transatlantic journeys by air transport rather than through sea routes. By 1960s, commercial aviation was transformed by the jet airliners which made air travel faster and more convenient. With the advent of the jet technology, for passengers, flying issues such as flight turbulence and engine noise within the cabin, that were present in earlier generations of the aircraft, were almost non-existent.

Gradually, the air travel option was becoming more and more accessible to the masses and the impact it had on the economy of the aviation industry was evident globally. To reduce the transatlantic flight time and to cater to a niche class of passengers, British Airways and Air France introduced the Concorde aircraft in 1976. Flying at a top speed of Mach 2, i.e., twice the speed of the sound, the Concorde significantly reduced the New York–Paris or New York–London flight time to three and a half hours. However, due to environmental concerns and high operational costs, Concorde ceased operations in 2003 (Grant, 2017, p. 393). During the 1980s, a significant achievement was recorded in the aviation history in 1986, when the first around-the-world non-stop flight on a Voyager aircraft was accomplished by Dick Rutan (b.1938) and Jeana Yeager (b.1952) in nine days. Meanwhile, in the last decade of the 20th century, air travel had become a means of connecting a world where more than a billion passengers were flying annually. To serve this growing consumer base eager to use air transportation, commercial airlines began focusing on cost cutting and compromising on safety aspects rather than considering speed as the determinant to enhance their market share.

1.5 2000 AND BEYOND: CHANGING LANDSCAPES

In the last century, major developments in the aviation operations were primarily recorded in Europe and the United States. Furthermore, due to infrastructural development and connectivity, the major concentration of the passenger traffic flow was also around these geographical regions. However, with globalization, growing population and the emergence of many economic hubs in Asia, such as China, Singapore, and Dubai, the focus of aviation operators has shifted toward the emerging Asian markets. For instance, in China alone more than 500 aircraft carried around 65 million passengers in 2000 which increased to 438 million (carried by 2,500 aircraft) in 2015 (Grant, 2017, p. 428). With the current growth and expansion of the aviation industry in China, in the next decade, i.e., by 2030s, China is expected to become the leader in terms of the scale of aviation operations and passenger volumes (Grant, 2017, p. 428). Besides the operational aspects and the swing in the revenue generation prospects toward Asian countries, there has been a paradigm shift in the global aviation industry – consideration for sustainable operations. There is a growing recognition among the aviation industry stakeholders regarding the adverse impact of the use of fossil fuels in aircraft operations. Industry stakeholders such as industry regulators, government bodies, environmental groups, and, in some parts of the world, passengers too, exert pressure on airline operators for environmental-friendly operations. For instance, Boeing in association with General Electric and Virgin Airlines had tested the first unmanned flight using biofuel in 2008 (IATA, 2009). Similarly, National Aviation Space Agency has carried out numerous research and development studies for low emission, more efficient aircraft. For further developments in aviation operations,

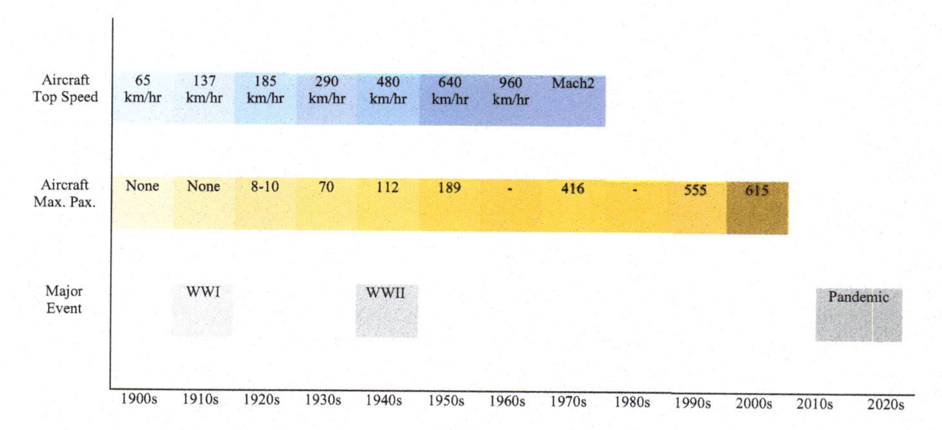

FIGURE 1.1 Timeline of major aspects in the civil aviation industry operations.

one of the major areas of focus is neither aircraft speed nor size, but fuel-efficient operations. Figure 1.1 captures this scenario and indicates that no gains have been made since 1970s to increase aircraft top speed. Similarly, as evident from Figure 1.1, since 1990s no significant increase in the aircraft passenger-carrying capacity has been observed. Considering the expected increase in the number of passengers (Pax) to more than a billion by mid-2030s and with increasing threats from greenhouse gas emissions, expansion, and the direction of aviation operations depends on the efforts of the concerned stakeholders to ensure sustainable aviation operations. Chapter 9 covers these aspects in detail.

In the 21st century, space tourism is a new dimension for commercial aviation operations. Unlike the aspirations and goals of aviation in the 20th century, the sky is no longer the limit for 21st century aviation. Space enthusiasts, business tycoons, and industrialists who have a passion for flying, spend millions of dollars to fulfil their desire of space travel. With the advancements in aviation technology, economic benefits, and a growing interest around commercial space travel, space travel is likely to become affordable for the masses before the end of this century.

1.6 CONCLUSION

Humanity's pursuit of flight began with flimsy kites and led to gigantic aircraft that mastered the skies. The drive to reach beyond the grasp of gravity has been one of the most dominant ones in human history and the story of aviation is full of accounts where human beings have even sacrificed their lives to fulfill the desire to fly. Over the span of a century, through numerous accounted and unaccounted for, individual and collective, efforts in different parts of the world, the art and science of flying has evolved beyond measure. Technological leaps and institutionalized developments have made the aviation industry operations remarkably fast, efficient, and safe. Present-day aircraft carry hundreds of passengers across the world, year-round, covering large distances in different and, at times, challenging weather conditions. However, the future of aviation industry operations depends on the decisions industry stakeholders make to ensure economically, environmentally, and socially sustainable aviation operations.

CHAPTER QUESTIONS

Q1. Discuss the progression of technological developments in civil aviation operations. List five most significant developments and justify their selection.

Q2. Discuss the progression of operational developments in civil aviation operations. List five most significant developments and justify their selection.

Q3. Analyze the impact of the transition from the developments made in the first half of the 20th century to the latter half in the civil aviation industry.

Q4. Analyze the impact of the transition from the developments made in the second half of the 20th century on the civil aviation operations of the first two decades of the 21st century.

Q5. Based on the trajectory of developments in the civil aviation industry, identify five most probable areas for future developments in the aviation operations. Justify your selection.

REFERENCES

Altuntas, O., Sohret, Y., & Karakoç, T. H. (2019). Fundamentals of sustainability. In T. H. Karakoç, C. O. Colpan, O. Altuntas & Y. Sohret (Eds), *Sustainable Aviation* (p. 122). Springer.

Anderson, Jr. J. D. (2002). *The Airplane: A History of Its Technology* (pp. 10–25). American Institute of Aeronautics and Astronautics: Reston, VA.

Eves, J. (2018). Who flew first? Gustave Whitehead and the Wright Brothers. *Connecticut History Review*, *57*(2), 185–189.

Grant, R.G. (2017). *Flight: The Complete History of Aviation*. DK Publishing.

IATA (2009). Annual General Meeting. www.iata.org/en/events/agm/2009/

ICAO (n.d.) History: The beginning www.icao.int/EURNAT/Pages/HISTORY/history_1910.aspx

2 Global Aviation Operations

CHAPTER OBJECTIVES

At the end of this chapter, you will be able to

- Differentiate between different international aviation regulatory bodies.
- Understand International Air Transport Association (IATA) traffic areas and global indicator system.
- Distinguish between airport categories.
- Identify major International Civil Aviation Organization (ICAO) and IATA airport codes.
- Know about major airlines operating internationally.
- Understand the structure of different airline networks.
- Get an overview of the impact of COVID-19 on the civil aviation industry.

2.1 INTRODUCTION

Over the past few decades, the burgeoning aviation industry has seen incredible advancements in technology, growth in passenger numbers, and an increase in economic hubs across the world, resulting in increased global mobility. Yet the primary requirement from air travel remains safety and speed and the aviation industry attempts to fulfill these needs in the most efficient and effective manner. Serving more than four billion travelers and shipping more than 60 million tons of freight annually (O'Connell & Bueno, 2018) the aviation industry is one of the fastest growing industries in the world, thanks to its affordability and convenience. It's a highly dynamic industry that operates under strict industry and government regulations, depends on state-of-the-art technology, faces intense competition, enforces stringent security measures, and struggles with high and fluctuating fuel costs (Riwo-Abudho et al. 2013; Stamolampros & Korfiatis 2019). And despite operational constraints, the aviation industry helps establish trading channels, promotes tourism, generates employment, and contributes toward economic development (ICAO, 2022; Tarkinsoy & Uyar, 2017). Resultantly the contribution of aviation operations in the global GDP is more than 3% or USD 2.5 trillion (Agrawal, 2020). Within the framework of the aviation industry, there are four different stakeholders interacting at different levels in

DOI: 10.1201/9780203731338-2

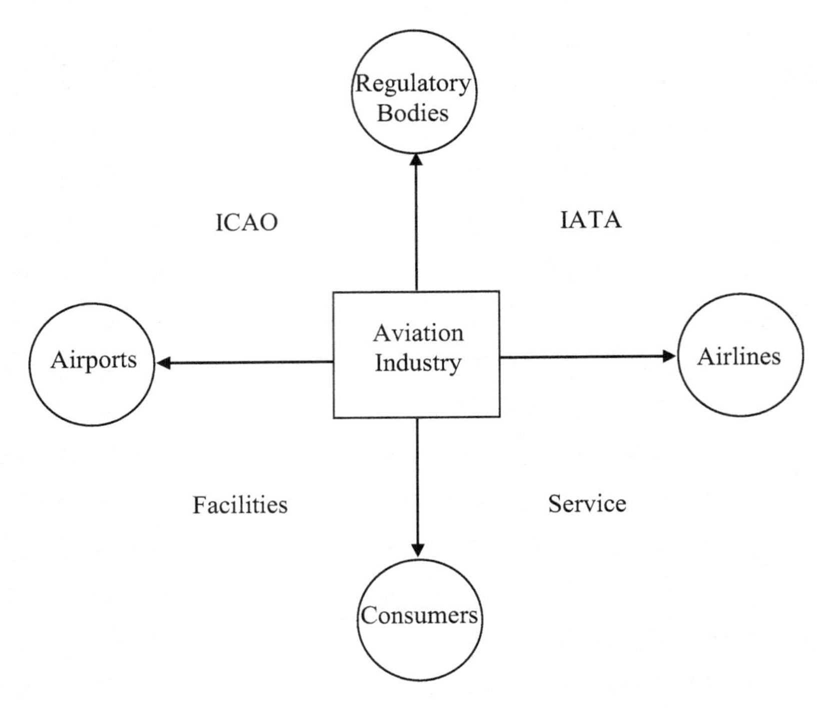

FIGURE 2.1 Major aviation operations' stakeholders.

the aviation supply chain, i.e., regulatory bodies, airports, airlines, and consumers. In Figure 2.1, the aviation industry supply chain stakeholders are presented.

2.2 AVIATION REGULATORY BODIES

With the expansion of air travel, airlines across the world operate to cater to a range of destinations to suit their business objectives. Therefore, to manage airline operations, monitoring the traffic and maintaining the logistical requirements within the industry, several accredited bodies function for the standardization and safe aviation operations globally. The primary role of these regulatory bodies is to establish, standardize, monitor, and implement the regulations for the industry stakeholders such as airlines and airports. For instance, ICAO formulates and establishes regulations for airport operations. Whereas the primary role of IATA is to establish standards for airlines. Aviation regulatory bodies update their standards based on the changing industry requirements as appropriate to ensure effective and efficient aviation operations.

2.2.1 International Civil Aviation Organization

ICAO is a United Nations body responsible for developing policies, international laws, and standards for civil aviation operations across the globe. ICAO was formed

TABLE 2.1
ICAO airport code structure

Letter Placeholder	Meaning
1st	Indicates continent or the region of the continent of the airport.
2nd	Indicates the country within the continent or the region.
3rd and 4th	Indicates the airport in the country.

at the Chicago Convention in 1944 to coordinate and regulate international air travel (ICAO, 1944). The vision of ICAO is to achieve sustainable growth across the industry by streamlining the regulations for its member states. There are 193 member states affiliated for whom ICAO establishes protocols for safety, air navigation capacity, security, economic development, environmental protection, and air accident investigation-related aspects (ICAO, n.d.). The information pertaining to these areas is published by ICAO through various articles and annexes. There are 96 articles laid out by ICAO and each article represents a rule that the member state is required to follow, whereas the standards and recommended practices are presented in 19 annexes. Through these articles and annexes, ICAO ensures that both airports and airlines function safely and efficiently through international cooperation of its member states. ICAO identifies the airports with a four-letter unique identifier based on its location. This unique airport identifier serves as a location indicator for the pilots and the air traffic control (ATC). Broadly, ICAO codes represent the location indicator based on the structure presented in Table 2.1.

However, there is an exception in the ICAO naming convention presented in Table 2.1. The airports based in the United States follow a different approach. For instance, ICAO airport code KATL represents Hartsfield–Jackson Atlanta International airport where the first letter 'K' indicates the identifier used for United States and the last three letters represent the airport. There is no regional identifier used for US airports in ICAO codes. Similar to the airport codes, ICAO also distinguishes between airlines through a specific three-letter code discussed later in this chapter.

2.2.2 International Air Transport Association

IATA is a trade association of airlines. There are 290 airlines based in 120 countries who are the members of IATA (IATA members, n.d.). One of the aims of IATA is to regulate air travel through appropriate pricing. Factors, such as crude oil prices, and resources, including labor and raw materials, are critical in establishing the operational costs of the airline which vary considerably in different parts of the world. Similar to ICAO, IATA has its own airport and airline naming systems. To manage global airline operations, depending on the geographic locations of the countries, IATA has categorized the countries in three major traffic conference areas (TCAs). For instance, countries in North America, South America, Central America, and the Caribbean are placed in TCA 1. Countries in Europe, Africa, and the Middle East fall

TABLE 2.2
IATA traffic conference areas

Hemisphere				
Western	Eastern			
TCA 1	TCA 2			TCA 3
North America	Europe	Africa	Middle East	Southeast Asia
Central America		Central Africa		South Asian Subcontinent
The Caribbean		Eastern Africa		Southwest Pacific
South America		Western Africa		Japan Korea
North Atlantic		Southern Africa		
South Atlantic		Central Ocean Islands		
Mid Atlantic		Libya		

under TCA 2. Along with Japan and Korea, countries which are in Southeast Asia, the South Asian sub-continent, and Oceania are part of TCA 3. An overview of the TCAs is presented in Table 2.2.

Furthermore, global indicators are also established by IATA to ascertain the flight routes which help the airlines in estimating the cost of travel.

2.2.2.1 Western Hemisphere (WH)

An itinerary with the port of origin and destination in the western hemisphere (TCA 1) is assigned with Western Hemisphere (WH) as its global indicator. For instance, ports A and B (Figure 2.2) are in the WH and if a flight departs from port A and arrives at port B or vice versa then this flight will have WH as its global indicator.

2.2.2.2 Eastern Hemisphere (EH)

Itineraries that originate and end in the eastern hemisphere (TCA 2 or TCA 3) are given Eastern Hemisphere (EH) as their global indicator. For instance, if a journey begins at port C and ends at port D (Figure 2.2) i.e, within TCA 2, then the journey will have EH as its global indicator. Similarly, EH global indicator is also assigned to flight routes which have origin (port E) and destination (port F) in TCA 3 (Figure 2.2). Furthermore, if a travel itinerary starts from any port in TCA 2 (port C or port D) and ends in any port in TCA 3 (port E or port F), then the global indicator for the itinerary will also be EH.

2.2.2.3 Pacific (PA)

Flights which fly over the Pacific Ocean between the port of origin and port of destination are assigned PA as the global indicator. Based on Figure 2.2, for instance if a flight departs from TCA 1 (port A or B) and flies across the Pacific to arrive at TCA 3 (port E or F) then PA will be the global indicator for this flight. Similarly, for flight itineraries starting from TCA 1 (port A or B) and ending in TCA 2 (port C or D) the

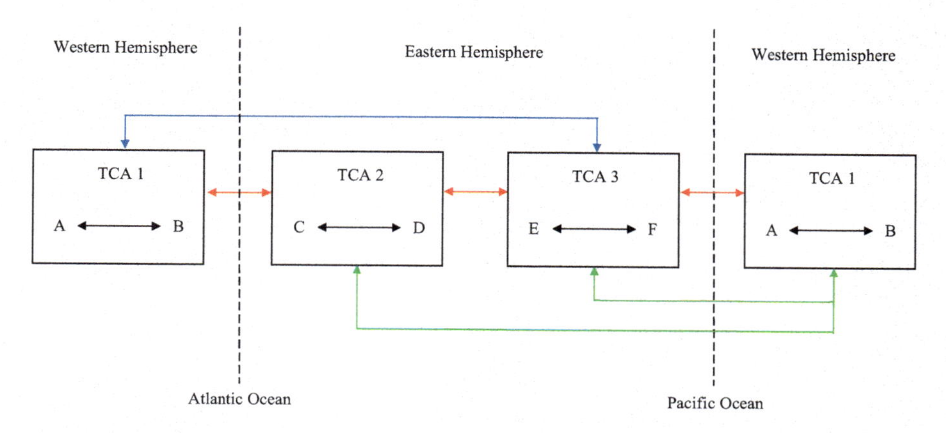

FIGURE 2.2 IATA traffic conference areas.

global indicator will be PA. However, in case an itinerary includes TCA 1 (port A or B) and TCA 3 (port E or F) across the Pacific via North America then Pacific North (PN) is assigned as the global indicator for the travel.

2.2.2.4 Atlantic (AT)

The itineraries that involve travel over the Atlantic Ocean i.e., destinations between TCA 1 (port A or B) and TCA 2 (port B or C) or TCA 1 (port A or B) and TCA 3 (port E or F) are assigned AT as the global identifier.

2.2.2.5 Atlantic/Pacific (AP)

For the flights that fly over both Atlantic and Pacific Oceans, the global indicator will be AP. For instance, if the travel includes TCA 2 (port C or D) to TCA 1 (port A or B) via any destination in North America will also be assigned AP as global indicator since both Atlantic and Pacific are crossed during the travel. However, there are some exceptions in assigning the global indicators such as SA, FE, RU, and TS.

2.2.2.6 South Asia (SA)

The travel itineraries that include flights from South Atlantic ports in TCA 1 to Southeast Asia in TCA 3 through southern African countries in TCA 2 will be assigned SA as the global indicator for the trip. It is important to know that the SA global indicator overrules AT global indicator. In this instance, even though the flight crosses the Atlantic Ocean but since a port in South Atlantic is a part of the itinerary, so SA as global indicator is used for the trip.

2.2.2.7 Far East (FE)

Russia is a large country which spans two continents: Europe and Asia. For global indicators, cities in Russia are demarcated accordingly. For travel itineraries involving Russian ports to another TCA 3 port on a non-stop flight, FE is used as the global indicator. In this instance, even though the entire journey is in the EH, EH can also be used as the global indicator. However, FE is assigned over EH as the global indicator for this itinerary.

2.2.2.8 Russia (RU)

In case of flights originating from European Russian ports to TCA 3 ports through Japan/Korea, RU will be the global indicator. In this instance as well, the entire journey is in the EH, so EH can also be used as global indicator. However, global indicator RU will be assigned for the trip over EH.

2.2.2.9 Trans-Siberian (TS)

In the case of flights operating between TCA 2 to TCA 3 through Japan/Korea, TS will be the global indicator for the itinerary.

Apart from generating airport and airline codes for the smooth functioning and efficient operations, IATA also uses different codes to identify airline schedules, easier baggage handling at the airports, ticketing, and reservation information for identifying passenger itineraries and cargo destinations. IATA achieves this using state-of-the-art technology and by exchanging real-time data with the stakeholders.

2.2.3 European Aviation Space Agency

European Aviation Space Agency (EASA) is an organization consisting of European countries for standardizing aviation safety standards and procedures for its member states (EASA members, n.d.; Appendix A). Major aspects of EASA's operations include aircrew licensing, air traffic controller operations, certifications, airworthiness, and airport regulations. For civil aviation operations, EASA cooperates with other international bodies such as ICAO and ensures that the regulations and standards of ICAO are maintained and practiced across its member states. One of the main objectives of EASA is to establish a single European sky to gain efficiencies for civil aviation operations in Europe.

2.2.4 Federal Aviation Administration

Federal Aviation Administration (FAA) is a federal agency in the United States that regulates all aspects of civil aviation operations. FAA is responsible for airport management, ATC, and certification of aircraft and personnel. The focus of FAA's operations is to ensure safety and efficiency using state-of-the-art technology across the United States. Among other aspects, FAA is responsible to develop airports to meet the increasing demands of the civil aviation sector through airport improvement programs (Atkinson, 2020). FAA also conducts training programs using simulations (virtual reality) through web-based training tools to minimize dependence on traditional instruction methods (Updegrove & Jafer, 2017).

2.3 AIRPORT LOCATIONS AND CATEGORIES

In this chapter, airports are classified into three broad categories: type A, type B, and type C. Airports which are connected to less than five direct destinations are considered as type A whereas type B airports cater to the requirements of passenger traffic connecting to a network of more than ten direct destinations. Similarly, airports which are connected to more than 50 direct destinations are referred to as type C airports. For distinction among the airports, in the maps (Flight connections, n.d.) presented in this chapter, type A-, B-, and C-airports are represented with solid red, yellow, and blue circles, respectively. Based on this classification, majority of the airports are type A followed by type B and type C. In the following sections, major airports situated in different TCAs indicating their location and classification (Flightconnections, n.d.) and the volume of passenger traffic in the ten busiest airports across different TCAs (airport codes, n.d.) is presented.

2.3.1 Western Hemisphere (TCA 1)

2.3.1.1 North and Central America

Airport locations of TCA 1 situated in North and Central America are presented in Figure 2.3 (Flightconnections, n.d.) indicating different categories of airports and their spread in the WH.

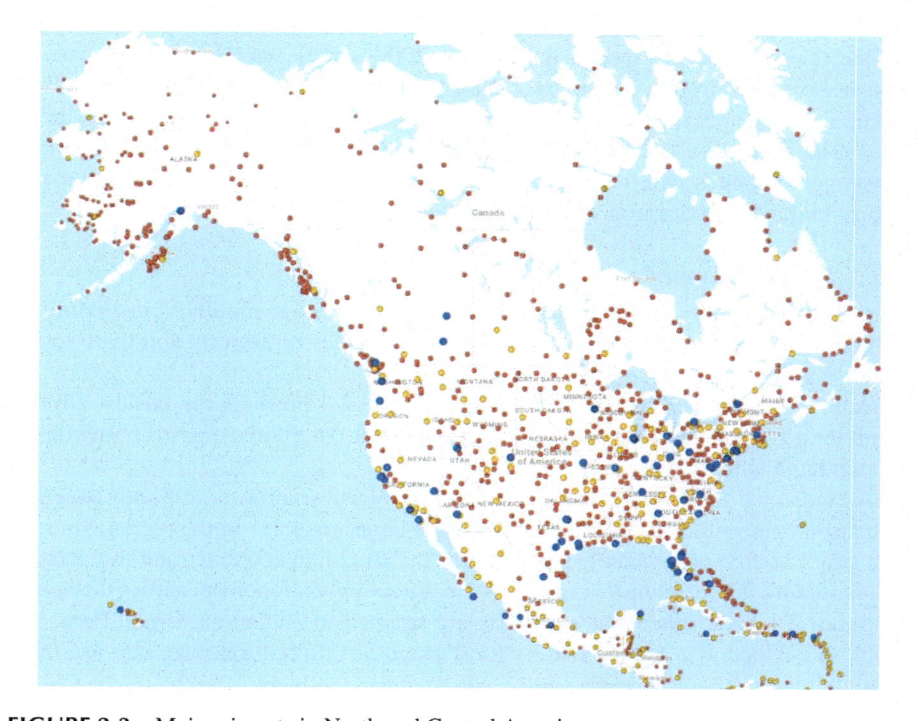

FIGURE 2.3 Major airports in North and Central America.

TABLE 2.3
Busiest airports in North America and Central America

Airport	Country	ICAO	IATA	Passengers
Hartsfield-Jackson Atlanta Int'l Airport	United States	KATL	ATL	107,394,029
Los Angeles International Airport	United States	KLAX	LAX	87,534,384
Chicago O'Hare International Airport	United States	KORD	ORD	83,245,472
Dallas Fort Worth International Airport	United States	KDFW	DFW	69,112,607
Denver International Airport	United States	KDEN	DEN	64,494,613
John F. Kennedy International Airport	United States	KJFK	JFK	61,623,756
San Francisco International Airport	United States	KSFO	SFO	57,738,840
McCarran International Airport	United States	KLAS	LAS	49,863,090
Seattle Tacoma International Airport	United States	KSEA	SEA	49,849,520
Toronto Pearson International Airport	Canada	CYYZ	YYZ	49,507,418

Among the airports shown in Figure 2.3, Table 2.3 presents the busiest airports (with their IATA and ICAO codes; airportcodes, n.d.) in North and Central America based on the inbound and outbound passenger traffic in 2018.

As evident from Table 2.3, except for one instance, i.e., Toronto Pearson International Airport in Canada, among the top ten busiest airports in North America, nine are situated in the United States. This is due to many factors such as; the deregulation of airlines in the United States which led to the emergence of many airline operators in the second half of the 20th century, easy connectivity to the vast network of airports across the country, severe competition among the airlines resulting in cheaper air fares for the passengers, improved aviation logistics, and so on. Furthermore, the tourism industry also played a significant role in encouraging passengers for air travel, which over the decades has become a norm rather than a luxury in the domestic circuit for travelers across the United States.

2.3.1.2 South America

Airport locations of TCA 1 situated in South America are presented in Figure 2.4 (Flightconnections, n.d.) indicating different categories of airports and their spread in the WH.

Among all airports shown in Figure 2.4, Table 2.4 presents the busiest airports (with their IATA and ICAO codes; airportcodes, n.d.) in South America based on the inbound and outbound passenger traffic in 2018.

As indicated in Table 2.4, among the top ten busiest airports in South America, five airports are situated in Brazil which is the largest country in the region. Tourism is a major source of revenue generation for the Brazilian economy and the aviation infrastructure in the country is conducive to attracting visitors from across the world. In South America, other major airports are situated in Argentina, Chile, Peru, and Columbia which also attract visitors for a range of different reasons who use these destinations as a point of entry and/or exit from the continent.

FIGURE 2.4 Major airports in South America.

TABLE 2.4
Busiest airports in South America

Airport	Country	ICAO	IATA	Passengers
São Paulo-Guarulhos International Airport	Brazil	SBGR	GRU	42,230,432
El Dorado International Airport	Colombia	SKBO	BOG	32,716,468
Jorge Chávez International Airport	Peru	SPJC	LIM	23,779,937
Comodoro Arturo Merino Benítez Int'l Airport	Chile	SCEL	SCL	23,324,306
São Paulo-Congonhas Airport	Brazil	SBSP	CGH	21,961,782
Brasília International Airport	Brazil	SBBR	BSB	17,855,163
Rio de Janeiro-Galeão International Airport	Brazil	SBGL	GIG	15,005,304
Jorge Newbery Airport	Argentina			13,365,290
Ministro Pistarini International Airport	Argentina			11,201,350
Tancredo Neves International Airport	Brazil			10,591,138

2.3.2 Eastern Hemisphere (TCA 2)

2.3.2.1 Europe

Airport locations of TCA 2 situated in Europe are presented in Figure 2.5 (Flightconnections, n.d.) indicating different categories of airports and their spread in the EH.

Among all airports shown in Figure 2.5, Table 2.5 presents the busiest airports (with their IATA and ICAO codes; airportcodes, n.d.) in Europe based on the inbound and outbound passenger traffic in 2018.

Aviation industry in Europe is significantly developed due to aviation conducive infrastructure and short flying durations between the European cities. Despite a

FIGURE 2.5 Major airports in Europe.

TABLE 2.5
Busiest airports in Europe

Airport	Country	ICAO	IATA	Passengers
London Heathrow Airport	United Kingdom	EGLL	LHR	80,100,311
Charles de Gaulle Airport	France	LFPG	CDG	72,229,723
Amsterdam Airport Schiphol	The Netherlands	EHAM	AMS	71,053,157
Frankfurt Main Airport	Germany	EDDF	FRA	69,510,269
Istanbul Ataturk Airport	Turkey	LTBA	IST	67,981,446
Adolfo Suarez Madrid-Barajas	Spain	LEMD	MAD	57,891,340
Barcelona-El Prat Airport	Spain	LEBL	BCN	50,172,457
Munich Airport	Germany	EDDM	MUC	46,253,623
London Gatwick Airport	United Kingdom	EGKK	LGW	46,081,327
Sheremetyevo International	Russia	UUEE	SVO	45,348,150

well-developed road and rail network across Europe, air travel is the preferred mode of transportation resulting in high passenger traffic for the airline operators. Furthermore, Europe functions as a gateway that connects North America and Asia: two continents with considerably large populations and ever-increasing passenger traffic. Significant number of air travelers from Asia use European destinations for transit purposes for trans-Atlantic flights. Similarly, passengers from North America consider Europe as a hub in their journey toward Asian destinations.

2.3.2.2 Africa and the Middle East

Airport locations of TCA 2 situated in Africa and the Middle East are presented in Figure 2.6 indicating different categories of airports and their spread in the EH.

Among all airports shown in Figure 2.6 (Flightconnections, n.d.), Table 2.6 presents the busiest airports (with their IATA and ICAO codes; airportcodes, n.d.) in Africa and the Middle East based on the inbound and outbound passenger traffic in 2018.

Based on Table 2.6, Dubai International Airport is the busiest airport in Africa and the Middle East combined. In the last couple of decades, Dubai has emerged as one of the major destinations for business and leisure globally. Furthermore, the geographical location of Dubai is also significant to connect flights to and from North America and Europe with Asia helping airlines in extending their flight networks with their partner airline operators. King Abdulaziz airport in Jeddah, Saudi Arabia is another major and significant airport in the region catering to the needs of travelers from all over the world who visit for the region for religious purposes. Other significant airports in the region are in South Africa and Egypt which are primarily considered as major global tourist destinations and attract millions of visitors annually.

FIGURE 2.6 Major airports in Africa and the Middle East.

TABLE 2.6
Busiest airports in Africa and in the Middle East

Airport	Country	ICAO	IATA	Passengers
Dubai International Airport	UAE	OMDB	DXB	89,149,387
King Abdulaziz Int'l Airport	Saudi Arabia	OEJN	JED	41,200,000
Hamad International Airport	Qatar	OTHH	DOH	35,400,000
King Khalid International Airport	Saudi Arabia	OERK	RUH	26,772,525
O.R. Tombo International Airport	South Africa	FAOR	JNB	21,231,510
Cairo International Airport	Egypt	HECA	CAI	15,010,501
Addis Ababa Bole Int'l Airport	Ethiopia	HAAB	ADD	12,143,938
Cape Town International Airport	South Africa	FACT	CPT	10,752,246
Mohammad V International Airport	Morocco	GMMN	CMN	9,748,567
Algiers Houari Boumediene Airport	Algeria	DAAG	ALG	7,900,000

2.3.3 Eastern Hemisphere (TCA 3)

2.3.3.1 South Asian Subcontinent

Airport locations of TCA 3 situated in the South Asian Subcontinent are presented in Figure 2.7 indicating different categories of airports and their spread in the EH.

Among all airports shown in Figure 2.7 (Flightconnections, n.d.), Table 2.7 presents the busiest airports (with their IATA and ICAO codes; airportcodes, n.d.) in south Asian subcontinent based on the inbound and outbound passenger traffic in 2018.

In Table 2.7, all the busiest airports in the south Asian subcontinent are in India. For this, one of the major reasons can be attributed to the surge in domestic air travel in India. Moreover, severe competition among the low-cost carriers has resulted in a significant drop in the air fares making air transport one of the preferred modes of transportation for the masses. Major international airports such as DEL, BOM, BLR, MAA, CCU, and HYD also cater to passengers from different international destinations making them hubs which are connected to other regional airports.

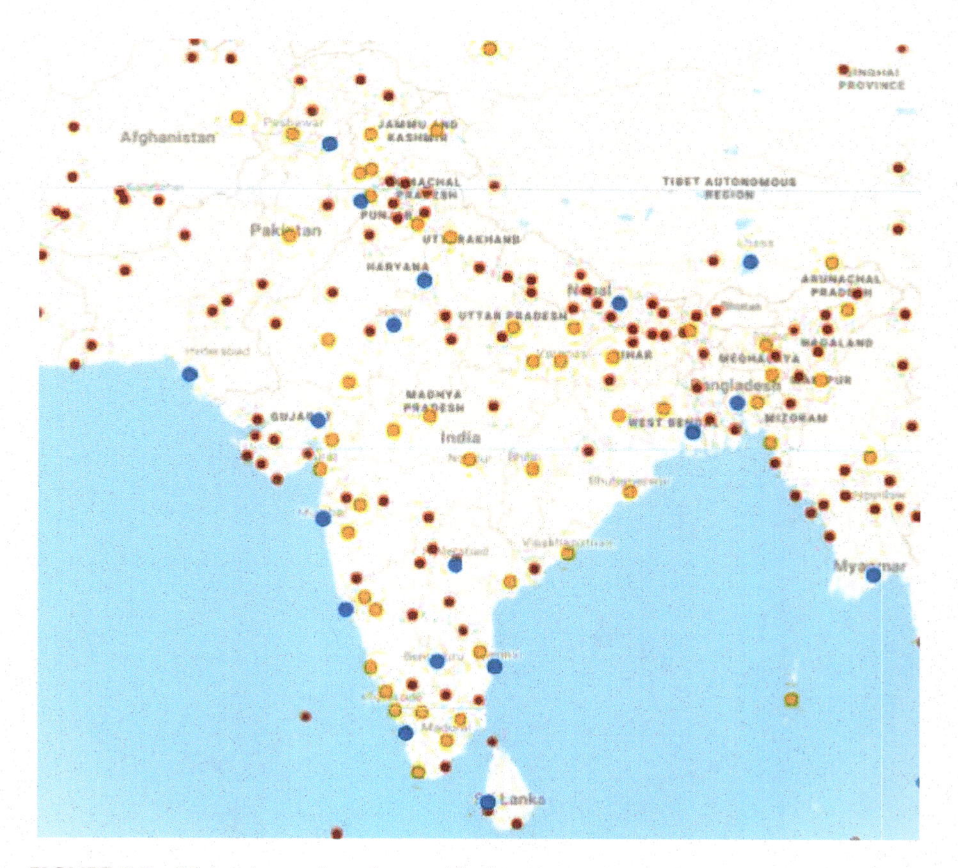

FIGURE 2.7 Major airports in and around India.

TABLE 2.7
Busiest airports in India

Airport	Country	ICAO	IATA	Passengers
Indira Gandhi International Airport	India	VIDP	DEL	69,866,994
Chhatrapati Shivaji Maharaj Int'l Airport	India	VABB	BOM	49,877,918
Kempegowda International Airport	India	VOBL	BLR	33,300,000
Chennai International Airport	India	VOMM	MAA	22,243,650
Netaji Subhas Chandra Bose Int'l Airport	India		CCU	22,015,391
Rajiv Gandhi International Airport	India		HYD	21,651,878
Sardar Vallabhbhai Patel Int'l Airport	India		AMD	11,432,996
Cochin International Airport	India		COK	9,624,334
Goa International Airport	India		GOI	8,356,240
Pune Airport	India		PNQ	8,085,607

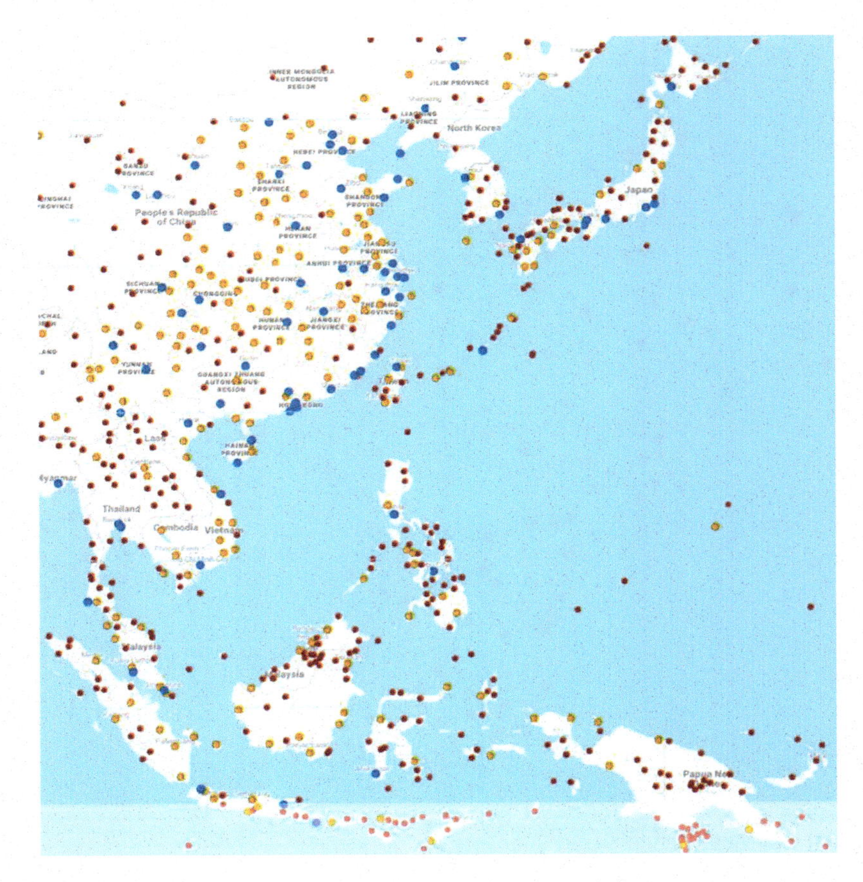

FIGURE 2.8 Airports in China, Japan, Korea, and Southeast Asia.

TABLE 2.8
Busiest airports in China, Japan, Korea, and Southeast Asia

Airport	Country	ICAO	IATA	Passengers
Beijing Capital International Airport	China	ZBAA	PEK	100,980,000
Tokyo Haneda International Airport	Japan	RJTT	HND	87,098,683
Hong Kong International Airport	Hong Kong	VHHH	HKG	74,700,000
Shanghai Pudong International Airport	China	ZSPD	PVG	74,050,000
Guangzhou Baiyun International Airport	China	ZGGG	CAN	69,730,000
Seoul Incheon International Airport	South Korea	RKSI	ICN	68,259,763
Soekarno Hatta International Airport	Indonesia	WIII	CGK	66,908,159
Singapore Changi Airport	Singapore	WSSS	SIN	65,600,000
Suvarnabhumi Airport	Thailand	VTBS	BKK	63,378,923
Kuala Lumpur International Airport	Malaysia	WMKK	KUL	59,959,000

2.3.3.2 Southeast Asia, China, Japan, and Korea (TCA 3)

Airport locations of TCA 3 situated in Southeast Asia, Japan, and Korea are presented in Figure 2.8 (Flightconnections, n.d.) indicating different categories of airports and their spread in the EH.

Among all airports shown in Figure 2.8, Table 2.8 presents the busiest airports (with their IATA and ICAO codes; airportcodes, n.d.) in Southeast Asia, China, Japan, and Korea based on the inbound and outbound passenger traffic in 2018.

Similar to European airports (Table 2.5), intense competition is observed among the airports in this region. As indicated in Table 2.8, airports in China (PEK, PVG, and CAN) compete with Tokyo (HND), Hong Kong (HKG), and Seoul (ICN) international airports. These airports majorly connect flights between Asia and North America (across the Pacific) and Europe.

2.3.3.3 Southwest Pacific (TCA 3)

Airport locations of TCA 3 situated in the Southwest Pacific are presented in Figure 2.9 (Flightconnections, n.d.) indicating different categories of airports and their spread in the EH.

Airports in Oceania are not as busy as the ones in other parts of the world. For instance, Sydney airport in Australia which is the busiest airport in the region catered to over 41 million passengers in 2018 (Table 2.9). Whereas, during the same period, Kuala Lumpur airport (KUL) in Malaysia which is the least busy airport among the top ten Asian airports served around 60 million passengers (Table 2.8).

Among all airports shown in Figure 2.9, Table 2.9 presents the busiest airports (with their IATA and ICAO codes; airportcodes, n.d.) in Southwest Pacific based on the inbound and outbound passenger traffic in 2018.

There are two major reasons for the significant difference in the passenger traffic in the two regions. First, the location of two airports. Malaysia is one of the major destinations in south-east Asia, primarily connecting the region with the rest of the world. Whereas, on the other hand, the Sydney airport, besides serving other

FIGURE 2.9 Major airports in Southwest Pacific.

TABLE 2.9
Busiest airports in Southwest Pacific

Airport	Country	ICAO	IATA	Passengers
Sydney Airport	Australia	YSSY	SYD	41,870,000
Melbourne International Airport	Australia	YMML	MEL	36,706,000
Brisbane International Airport	Australia	YBBN	BNE	23,205,702
Auckland International Airport	New Zealand	NZAA	AKL	20,025,922
Daniel Inouye International Airport	United States (Hawaii)	PHNL	HNL	19,950,707
Perth International Airport	Australia	YPPH	PER	13,690,610
Adelaide International Airport	Australia	YPAD	ADL	8,090,000
Christchurch International Airport	New Zealand	NZCH	CHC	6,566,598
Gold Coast Airport	Australia	YBCG	OOL	6,457,086
Kahului Airport	United States (Hawaii)	PHOG	OGG	6,444,000

destinations, mostly serves as the hub for the trans-Pacific and Asia bound flights. Second, the difference in the population in and around the two regions, which varies significantly, impacts passenger volumes considerably. Major airports in the south pacific region are in Australia (SYD and MEL) and New Zealand (AKL and CHC).

In summary, the aviation industry in the developing world, and especially in Asia, is witnessing a big thrust in airline operations. As evident from Table 2.8, the total number of passengers served across the ten busiest airports in Asia were more than the number of passengers served in the ten busiest airports in North America (Table 2.3) and Europe (Table 2.5). Countries such as China and India are among the largest and fastest growing markets for air travel. A major reason for high passenger traffic in Asia can be ascribed to the population of Asian countries and infrastructural development. Also, in the last few decades, airports such as Dubai International (DXB) and Guangzhou Baiyun International (CAN) serve as major hubs in connecting flights to and from Europe, the Middle East, North America, and Oceania.

2.4 AIRLINES

Airlines operate worldwide. Many airline operators including Full-Service Carriers and Low-Cost Carriers serve domestic and international destinations globally. In terms of fleet size, the airline operators based in the United States (American Airlines, Delta Airlines, Southwest Airlines, and United Airlines) have the largest fleet of aircraft. Also, some airline carriers in Europe (Ryan Air and IAG) and China (China Southern, China Eastern, and Air China) also have more than 400 aircraft in their fleet, indicating the scale of their operations. Similarly, in terms of annual passenger traffic, United States-based airline carriers are the largest followed by Europe and China-based airline operators. Table 2.10 presents the largest airlines (airportcodes, n.d.) with regards to fleet size and the volume of passengers carried annually.

Table 2.10 indicates the dominance of some airline operators in the civil aviation industry. However, it is important to know that not all destinations are directly served by these operators in their international operations for which they use code sharing to expand their reach to wider markets to suit their business model. Airlines operate their flight network either through hub-and-spoke (H&S) or point-to-point network operations as discussed in the following sections.

2.4.1 HUB-AND-SPOKE NETWORK

In the H&S network, unless a hub is the destination, the hub airport is used as a transit point to get to another destination which is not directly connected with another port

TABLE 2.10
Ranking of largest airlines in terms of fleet size and passenger volume

		Fleet size		Passenger Traffic		CODE	
Airline	**Country**	**Count**	**Rank**	**Count**	**Rank**	**IATA**	**ICAO**
American Airlines	United States	956	1	199,600,000	1	AA	AAL
Delta Airlines	United States	879	2	186,400,000	2	DL	DAL
Southwest Airlines	United States	749	4	157,800,000	3	WN	SWA
United Airlines	United States	765	3	148,100,000	4	UA	UAL
Ryanair	Ireland	439	8	130,300,000	5	FR	RYR
Lufthansa	Germany	338	10	130,000,000	6	LH	DLH
China Southern Airlines	China	597	6	126,300,000	7	CZ	CSN
China Eastern Airlines	China	525	7	110,800,000	8	MU	CES
IAG	Spain/UK	598	5	104,800,000	9		
Air China	China	418	9	101,600,000	10	CA	CCA

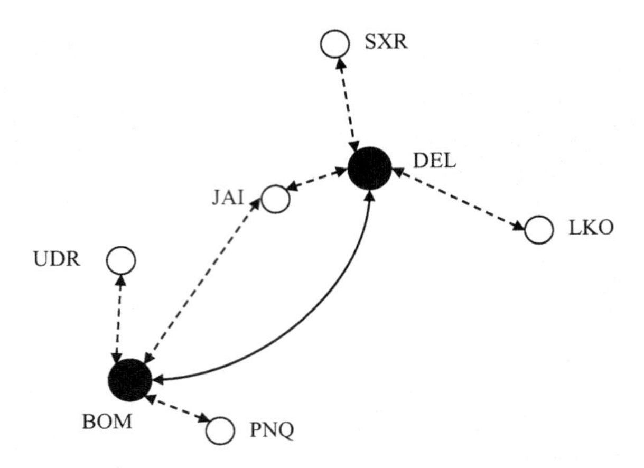

FIGURE 2.10 A hub-and-spoke network.

the airline serves. Majority of the passengers use a hub only as a point of connection between flights and not as a port of origin and/or destination. The H&S network caters to many destinations and covers the market in a wider geographical area. In this network, multiple aircraft fleets are used to serve the destinations by providing daily flight services between the city-hub pair and schedules are optimized with other sectors for the connecting passengers. To achieve this, H&S network functions with the optimal utilization of the assets such as aircraft and crews to minimize cost and gain efficiencies in the operations for the airline carrier. For instance, based on Figure 2.10, Delhi (DEL) is a hub which connects the flights from Srinagar (SXR), Lucknow (LKO), and Jaipur (JAI). Similarly, another hub (BOM) connects the flights from Udaipur (UDR), Pune (PNQ), and JAI. For a passenger who plans a trip from UDR to SXR it is required to get to BOM (a hub) and get the connecting flight to another hub (DEL) to reach SXR.

Similarly, to get to PNQ from LKO, a flight from LKO to DEL and DEL to BOM is required before proceeding on the final leg of the journey from BOM to PNQ. The tight scheduling of resources results in small time windows (to wash out minor delays in the schedule) which may significantly affect the planned operations during schedule perturbations, if any.

2.4.2 Point-to-Point Network

In a traditional (or pure) point-to-point (P2P) network (Figure 2.11), passengers board a flight at the port of origin and reach their destination through a non-stop flight. In this system, the intermediary hubs between the origin and destination are not involved. In a P2P network, each route is independent from another. Airlines which use this network, serve to a relatively smaller market using a single aircraft fleet to maximize the aircraft capacity in each sector. The frequency of flights in each route varies depending on passenger demand. In this network, for passengers, the total travel duration is less due to the direct access to the destination.

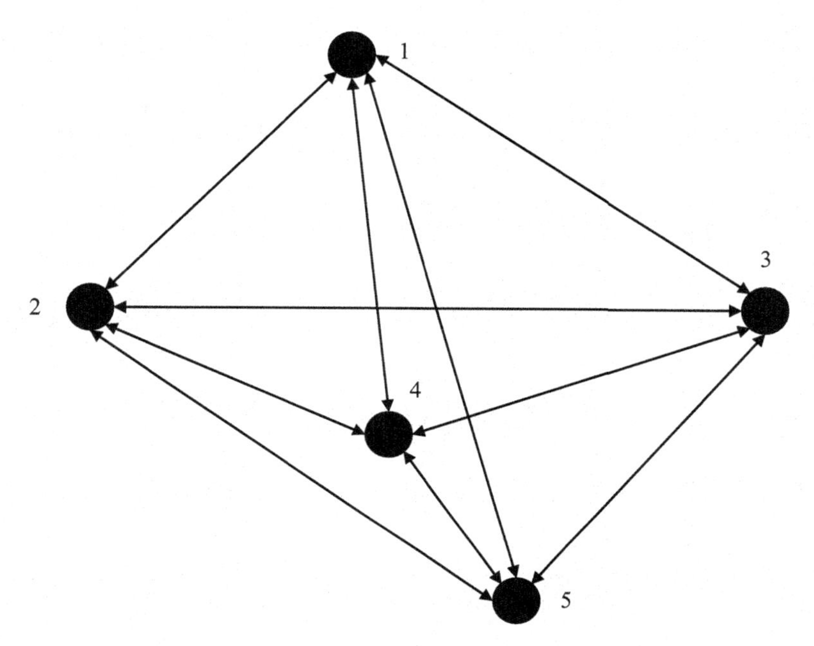

FIGURE 2.11 A point-to-point network.

In P2P networks, during disruptions, the ripple effect of the disruption is insignificant as compared to the effect of similar disruption at a hub port in the H&S network. Based on the target market and business model, present-day airlines use either H&S or P2P networks in their operations. Generally, large airlines such as Emirates, British Airways, Lufthansa, and so on use the H&S network whereas the operations of small airline carriers including Ryanair and Air Asia are based on P2P networks. Overall, the costs of functioning with the H&S are high due to the volume of passenger traffic, gate allocation for incoming outbound flights, logistics required in baggage handling, and maintenance of resources.

Almost all airlines operate with multiple fleets consisting of a range of aircraft types which vary in size and seating capacity. Airlines schedule their flights using the appropriate fleet-type based on their network configuration (point-to-point or H&S) depending on the frequency of the operations and demand (Janić, 2014).

2.5 CONSUMERS

Information technology and the Internet have exposed consumers to transparent and global markets. In the aviation industry, consumers such as air travelers/passengers and freight operators use the Internet to avail facilities in search for lower prices, shorter lead times, and better service using a range of technological mechanisms. A majority of the consumers are passengers and based on the reason of their travel (leisure or work), passengers travel in groups or as solo travelers. Alternatively, based on their travel objectives, passengers can also be classified as time-sensitive, cost-sensitive, and aviation lovers (Harrison et al., 2015). Regardless of passenger classification, all air travelers avail a range of aviation industry facilities and services resulting in different experiences to the service providers and consumers (Bunchongchit & Wattanacharoensil 2021; Fodness & Murray, 2007). Furthermore, consumers in the aviation industry not only demand quality but also perceived benefits from the products and/or services they expect from the aviation operations. Through this, consumers seek value against the price they pay for the goods and/or service. However, value is a broad concept and has another subjective dimension in non-monetized value. This aspect is not a direct measure of the monetary value and varies depending on the product or service a consumer receives.

2.6 IMPACT OF COVID-19 ON THE AVIATION INDUSTRY

Within the 21st century alone, the aviation industry has been affected by major setbacks such as the tragic events at the World Trade Centre in 2001, global recession in 2008, Iceland volcanic ash in 2010, spread of viruses such as Severe Acute Respiratory Syndrome in 2003, Swine flu in 2010, Middle East Respiratory Syndrome in 2012, Ebola in 2015 and Zika in 2016. However, the impact of these events was largely confined to a region and was relatively short-term (Drljača et al., 2020). Contrarily, the impact and ripple effects of the COVID-19 pandemic are unprecedented in the history of the aviation industry.

Due to the impact of COVID-19 pandemic, global travel restrictions affected the world population significantly, resulting in a monetary loss of more than USD 100 billion – the highest revenue deficit in the aviation industry to date (IATA, 2020; ICAO, n.d.). Similarly, in terms of the decline in the passenger volumes globally, the effect of the pandemic has been considerable (ICAO, 2021). During the peak of the pandemic in 2020, the decrease in air passenger traffic was observed to be less than 95% as compared to the passenger volume in 2019 and almost a complete standstill of the aviation traffic was observed when more than two-thirds of all commercial airliners were grounded (Kotoky et al., 2020; Drljača et al., 2020). However, in some instances, domestic flights and freight operations were functional with severe restrictions. Due to the impact of COVID-19 in 2020 and 2021, the significance of aviation operations in the development of economy indicated high dependency of the passenger movement and trade on the aviation industry operations worldwide.

Similar to the global aviation industry, the impact of COVID-19 on the aviation operations in India too has been significant, incurring considerable losses to airline operators and allied services (Agrawal, 2020). In India, aviation operations were completely suspended between March 2020 to May 2020 resulting in a revenue loss to the aviation sector in excess of INR 200 billion (Agrawal, 2020). The decline in air travel due to limited business and leisure travel is significant and the recovery of the aviation sector from the effect of the pandemic to pre-pandemic, i.e., 2019 operations is highly unlikely in the next five years (Pierce, 2020).

This leads to unprecedented challenges not only for airport and airline operations but also for the global aviation industry and opens a window of opportunity to explore new options and/or alternatives to make the aviation industry more resilient to combating such crises in the future. Attempts have been made to ensure COVID-19-free air travel and to prevent its spread, such as the Polymerase Chain Reaction (PCR) tests which have been introduced at the airports and elsewhere for COVID-19 detection (Tabares, 2021). For this, IATA classifies flights into three major categories, i.e., low-risk, medium-risk, and high-risk depending on the departure port of the flight, COVID-19 outbreak situation at the departure port, whether the aircraft is equipped with particulate air filters, and the duration of the flight (Nakamura & Managi, 2020). Based on the flight risk level, prevention measures are enforced at the airports where temperature checks and COVID-19 test results at the departure and arrival ports are carried out before letting passengers proceed with their journey. PCR testing at airports is either free of charge or costs between USD 75 to USD 225 depending on the time it takes to get the outcome of the test which ranges between 1 h and 72 h (Tabares, 2021). In case of detecting a positive COVID-19-infected passenger, it is mandatory for the passenger to self-isolate themselves or they are sent to a government-managed quarantine facility (Tabares, 2021).

Apart from introducing stringent industry regulations on air travel, providing health screening mechanisms and management of quarantine operations, making use of the technological advances is imperative to achieve the pre-pandemic travel experience. In this regard, the application of Industry 4.0 can be critical in developing the infrastructure using state-of-the-art technologies for aviation industry operations. This can be achieved by updating the conventional aviation operations such as immigration services (biometrics, digital documents, e-gates) and baggage handling which are time-consuming processes and, in many instances, require physical interaction with the airport systems (Drljaca et al., 2020). Through Industry 4.0, smart or contactless airport operations not only mitigate the risk of infections and physical interactions with the airport personnel but also enhance passenger safety, speed up the processes and provide better service quality (Drljaca et al., 2020).

Due to the significant effects of the COVID-19 pandemic, both airport facilities and airline operators introduced precautionary measures. To ensure increased cleanliness at the airport operations and during air travel, airports and airlines have taken measures to protect their personnel working at the airport and passengers through social distancing, enhanced sanitization, and by wearing masks before boarding, during the flight, and after deplaning the aircraft (Tabares, 2021).

2.7 CONCLUSION

The global civil aviation landscape has changed considerably over the past five decades. Owing to globalization and technological advancements, the aviation industry has evolved significantly and now plays a pivotal role within the global economy. Within the aviation industry there are a number of regulatory bodies that operate to implement policies and practices for airport and airline operations to achieve safety, efficiency, and convenience for the stakeholders. The global civil aviation operations are divided into TCAs based on the geographic locations of member countries. Airline operators serve numerous domestic and international destinations through their flight networks (H&S or point-to-point) spread across different TCAs. Recently, the impact of the COVID-19 pandemic on civil aviation operations has reverberated throughout the industry and it is now required to reflect upon the changed aviation scenarios through state-of-the-art mechanisms and address the gaps and shortcomings of existing operations.

CHAPTER QUESTIONS

Q1. Describe and discuss the role, scope, and significance of international aviation bodies and their impact on the global civil aviation operations.

Q2. Why global indicators are used in international civil aviation operations? Discuss two benefits and limitations of the global indicator system.

Q3. Comment on TCA 1, TCA 2, and TCA 3 with regards to the passenger traffic and list five busiest destinations in each TCA.

Q4. Discuss the difference between H&S and point-to-point networks. Identify two benefits and limitations of each from the perspective of airline operators.

Q5. Analyze the impact of COVID-19 pandemic on global aviation operations. Suggest two areas for improvement in the airport and airline operations to combat the effects of the pandemic.

REFERENCES

Agrawal, A. (2020). Sustainability of airlines in India with COVID-19: Challenges ahead and possible way-outs. *Journal of Revenue and Pricing Management*, *20*, 457–472.

Airport codes. (n.d.). Top 10 busiest airports https://airportcodes.io/en/?s=busiest+airports

Airport codes. (n.d.). Top 20 biggest airlines by fleet size https://airportcodes.io/en/blog/top-20-biggest-airlines-by-fleet-size/

Atkinson, C. L. (2020). The Federal Aviation Administration Airport Improvement Program: Who benefits? *Public Organization Review*, *20*(4), 789–805.

Bunchongchit, K., & Wattanacharoensil, W. (2021). Data analytics of Skytrax's airport review and ratings: Views of airport quality by passengers types. *Research in Transportation Business & Management*, *41*, 100688.

Drljača, M., Štimac, I., Bračić, M., & Petar, S. (2020). The role and influence of Industry 4.0 in airport operations in the context of COVID-19. *Sustainability*, *12*, 10614.

EASA. (n.d.). EASA member states www.easa.europa.eu/light/topics/easa-member-states

Flight connections. (n.d.). https://flightconnections.com FlightConnections

Fodness, D., & Murray, B. (2007). Passengers' expectations of airport service quality. *Journal of Services Marketing*, *21*(7), 492–506.

Harrison, A., Popovic, V., & Kraal, B. (2015). A new model for airport passenger segmentation. *Journal of Vacation Marketing, 21*(3), 237–250.

IATA. (n.d.). IATA Members www.iata.org/en/about/members/

ICAO. (n.d.). About ICAO. www.icao.int/about-icao/Pages/default.aspx

ICAO. (n.d.). Economic Impacts of Covid-19 on Civil Aviation. www.icao.int/sustainability/Pages/Economic-Impacts-of-COVID-19.aspx

ICAO. (1944). Convention of international Civil aviation – document 7300. www.icao.int/publications/Documents/7300_orig.pdf

ICAO. (2020). Economic Development Air Transport Monitor www.icao.int/sustainability/Documents/MonthlyMonitor-2020/MonthlyMonitor_June_2020.pdf

ICAO. (2021). 2020 passenger totals drop 60 percent as COVID-19 assault on international mobility continues. www.icao.int/Newsroom/Pages/2020-passenger-totals-drop-60-percent-as-COVID19-assault-on-international-mobility-continues.aspx

ICAO. (2022). Effects of COVID-19 on Civil aviation: Economic impact analysis. www.icao.int/sustainability/Documents/COVID-19/ICAO_Coronavirus_Econ_Impact.pdf

Janić M. (2014). Modelling the effects of different air traffic control (ATC) operational procedures, separation rules, and service priority disciplines on runway landing capacity. *Journal of Advanced Transportation*, *48*(6), 556–574.

Kotoky, A., Stringer, D., & Saxena, R. (2020). Two-Thirds of the World's Passengers Jets Are Grounded Amid COVID-19 Pandemic. https://time.com/5823395/grounded-planes-coronavirus-storage

Nakamura, H., & Managi, S. (2020). Airport risk of importation and exportation of the COVID-19 pandemic. *Transport Policy*, *96*, 40–47.

O'Connell, J. F. & Bueno, O. E. (2018). A study into the hub performance Emirates, Etihad Airways and Qatar Airways and their competitive position against the major European hubbing airlines. Journal of Air Transport Management, 69, 257–268.

Pierce, B. (2020). COVID-19, Outlook for air travel in the next 5 years. www.iata.org/en/iata-repository/publications/economic-reports/covid-19-outlook-for-air-travel-in-the-next-5-years/

Riwo-Abudho, M., Njanja, L. W., & Ochieng, I. (2013). Key success factors in airlines: Overcoming the challenges. *European Journal of Business and Management*, *5*(3), 84–88.

Stamolampros, P., & Korfiatis, N. (2019). Airline service quality and economic factors: An ARDL approach on US airlines. *Journal of Air Transport Management*, *77*, 24–31.

Tabares, D. A. (2021). An airport operations proposal for a pandemic-free air travel. *Journal of Air Transport Management*, *90*, 101943.

Tarkinsoy, J., & Uyar, A. (2017). Sustainability reporting in the airline industry. In K. Çalıyurt & Ü. Yüksel (Eds), *Sustainability and Management: An International Perspective* (pp. 100–118), Routledge.

Updegrove, J. A., & Jafer, S. (2017). Optimization of air traffic control training at the Federal Aviation Administration Academy. *Aerospace*, *4*(50) , 1–12.

3 Civil Aviation Landscape in India

CHAPTER OBJECTIVES

At the end of this chapter, you will be able to

- Understand the structure and functions of civil aviation regulatory bodies in India.
- Differentiate between different types of airports operational in India.
- Identify airport and airline codes used by International Air Transport Association (IATA) and International Civil Aviation Organization (ICAO).
- Distinguish between the operational statuses of the airports in India.
- Know about the airline operators functional in India.
- Get an overview of the developments in the Indian aviation industry.

3.1 INTRODUCTION

In the global civil aviation industry, India is an emerging market. The International Air Transport Association (IATA) states that India is among the largest and fastest growing markets for air travel with increasing growth in the domestic segment (IATA, 2018). The domestic air travel market in India is approaching 100 million passengers annually, and by 2026 it is expected to be the third-largest aviation market after China and the United States; and the largest by 2030 (IATA, 2018; IBEF, 2021). The aviation industry in India directly and indirectly provides more than 13 million jobs and contributes USD 30 billion toward India's GDP annually (IATA, 2018). Statistical data also indicates that the workforce in India will increase to 900 million by 2030 for whom the preferred mode of transportation for business and leisure is expected to be air travel (IBEF, 2017). Especially, with deregulation in the aviation industry in India and infrastructural investment by governmental institutions, a window of opportunity lies ahead for airline operators to tap into the burgeoning aviation market of the country. Therefore, in this chapter, numerous aspects are presented to provide an overarching view of the scope and limitations surrounding aviation operations. For

DOI: 10.1201/9780203731338-3

this, the role and significance of different civil-aviation-related regulatory bodies are presented to provide an overview of the civil aviation operations.

3.2 CIVIL AVIATION REGULATORY BODIES

Directorate General of Civil Aviation (DGCA) is a major regulatory body responsible for air transport services and safety aspects related to the aviation sector in India. DGCA falls under the Ministry of Civil Aviation and responsible for the regulation of civil aviation transport through its various directorates and divisions (DGCA Organisational structure, 2021; Figure 3.1). An overarching view of the scope of the directorates and divisions of DGCA is presented in the following sections to gain deeper understanding of numerous aspects related with the civil aviation operations.

3.2.1 Continuing Airworthiness

Airworthiness directorate is responsible for developing regulations, policies, and guidelines to ensure the airworthiness of an aircraft. Also, approvals and certifications of aircraft registration, renewals, and special flight permits are granted by the directorate. The regional and sub-regional offices of the directorate are responsible for the implementation of national aviation regulations, monitoring, and compliance of the approved maintenance records.

FIGURE 3.1 DGCA directorates and divisions.

3.2.2 Aircraft Certification

For aircraft certification, four divisions of the Aircraft Engineering Directorate cover different aspects of operations.

- The first division is Aero Engineering Division which is responsible for issuance, validation, and recognition of different types of aeronautical certificates.
- The second division, i.e., Aero Laboratories Division consists of various laboratories to address failure analysis, flight recorder, physical, chemical, and material aspects involved with the aviation operations.
- Aviation Environment Unit is the third division which disseminates the guidelines to develop and monitor carbon footprint, air quality of airports, and noise contours around airports as prescribed by regulatory bodies such as ICAO, Central Pollution Control Board and Airports Authority of India (AAI).
- Air Transport Division (ATD) is the fourth division which is responsible for the approval of flight schedules of the domestic airlines and Indian air carriers operating on international sectors. Scrutiny of airline and passenger traffic data regarding airline operations, engineering, safety, and ground facilities are the major aspects covered by the ATD.

3.2.3 Air Navigation Services

In collaboration with International Civil Aviation Organization (ICAO), the Air Navigation Services directorate ensures aviation safety for matters relating to air space management, i.e., flight route structure and flight control zones. Risks and safety hazards related to air traffic management are addressed by risk mitigation plans, following inspection schedules, and developing training programs.

3.2.4 Aircraft Operations

The flight standards directorate provides technical support to flight operations and authorizes instructors and line training captains. Furthermore, the directorate approves flight simulation devices and surveillance of air operators for the safe operation of the aircraft.

3.2.5 Legal Affairs

Legal affairs fall under the directorate of Information and Regulation. The focus of the directorate is to negotiate aviation agreements with foreign-based airlines, co-ordination of operations with ICAO, and to advise legalities to other directorates as appropriate.

3.2.6 Personal Licensing

Personal licensing directorate covers three main licensing aspects involving pilots, aircraft maintenance engineers (AME), and flight dispatchers. Pilot licensing includes flying training, AME covers airworthiness, and the flight despatcher ensures flight standards. The directorate issues grant extensions and renews licenses for the stated categories.

3.2.7 Aerodromes and Ground Aids

Aerodromes and Ground Aids directorate is responsible for aerodrome licensing, regulation of work at airports, and safety. Aerodrome standards and practices regarding operational procedures and infrastructure development, including aerodrome licensing, are considered by the directorate.

3.2.8 Administration

Administration captures the matters related to regional and sub-regional offices catering to aircraft engineering and transport. Aspects related to operational functionalities such as maintenance, recruitment, finance, vigilance, and coordination with other government institutions are addressed through a range of administrative tasks for the effective functioning of the departments involved with the civil aviation industry.

3.2.9 Flying Training and Sports

The directorate of Flying Training and Sports approves flying training organizations, chief flying instructors, and chief ground instructors. The directorate also permits the import and acquisition of aircraft for airlines based in India.

3.2.10 Air Operator Certification and Management

Air Transport directorate endorses permits and certificates for import and acquisition of aircraft. Domestic flight schedules and flight movements involving cargo flights, air ambulances, and charter flights fall under the purview of air transport directorate. DGCA heads the Air Operator Certification and Management Bureau for the issuance of air operator certificates and related management aspects.

3.2.11 Investigation and Prevention

Air safety directorate is responsible for investigation of civil aircraft incidents. Furthermore, the directorate carries out regular audits of airlines, issues civil aviation requirements, air safety circulars, and conducts spot checks and inspections. The directorate also approves flight safety manuals and coordinates with ICAO and other aviation bodies concerning aviation safety.

Under DGCA directorates, five divisions operate covering a range of civil aviation operations as presented below.

3.2.12 Information Technology

Information Technology caters to the requirements of DGCA regarding real time data exchange and other information through a range of databases acquired from various directorates.

3.2.13 Training

Training involves conducting training for DGCA and Indian Aviation Academy officers to issue appropriate credentials based on their accomplishments.

3.2.14 International Cooperation

To ensure the implementation of the policies of international regulatory bodies and enhancing safety of civil aviation operations in India, DGCA established the International Cooperation Group to address matters in relation to policy and technical aspects.

3.2.15 Surveillance and Enforcement

Surveillance and enforcement play a key role in planning, monitoring, and enforcing actions in maintaining databases of different directorates of DGCA.

3.2.16 State Safety Program

The state safety program carries out responsibilities pertaining to preparation, co-ordination, monitoring, and implementation of aviation safety programs.

DGCA has headquarters in New Delhi and operates nationwide with its regional offices located in different cities across India which are responsible for specific civil aviation functions (DGCA Regional directorates, 2021). The regional offices are located nationwide in four zones: northern, western, eastern, and southern. Each zone is responsible for specific functions based on the requirements and capacity of the directorate. One of the most important aspects of DGCA's operations is to comply with the regulations and standards of ICAO. As evident from Table 3.1, flight operations in India are controlled by the Western regional office. Similarly, aircraft engineering-related aspects are only confined to the Southern regional office indicating the expertise available at these centers. On the contrary, continuing air-worthiness is mandatory for regional offices to ensure safe and efficient civil aviation operations across India.

Based on different aspects of DGCAs operations, a summary of the functions of regional and other offices is presented in Table 3.1.

3.3 AIRPORTS IN INDIA

In India, there are more than 100 operational airports (34 international) that operate with scheduled commercial flights in the domestic and international sectors. Also, there are more than 25 civilian airport enclaves based at defense airbases across India catering to the regional demand. Regardless of the airport type, all airports are assigned an airport code by IATA and ICAO. The airport code is a unique identifier which helps airline operators and international regulatory bodies in identifying the airport for operational reasons.

TABLE 3.1
DGCA directorates and their major functions across different cities

Region	Cities	Flight Operations	Aerodrome Standards	Continuing Airworthiness	Air Safety	Aircraft Engineering	Training & Licencing
Northern	Delhi		✓	✓	✓		✓
	Lucknow			✓			
	Kanpur			✓			
	Patiala			✓			
Western	Mumbai	✓	✓	✓	✓		✓
	Bhopal			✓			
Eastern	Kolkata		✓	✓	✓		✓
	Patna			✓			
	Bhubaneshwar			✓			
	Guwahati			✓			
Southern	Chennai		✓	✓	✓		✓
	Hyderabad			✓	✓		
	Bengaluru			✓		✓	
	Kochi			✓			

3.3.1 Domestic Airports

India has a dense network of domestic airports connected with regional and international airports providing access to various destinations to passengers. In this chapter, airports in India are classified into three broad categories: type A, type B, and type C. The airports which are connected to less than five direct destinations are considered as type A whereas type B airports cater to the requirements of passenger traffic to a network of more than ten direct destinations. Similarly, airports which are connected with more than 50 direct destinations are referred to as type C airports. For distinction in airport types, type A-, B-, and C-airports are represented with red, yellow, and blue circles, respectively, in the maps presented in this chapter (Flight connections, n.d.).

For instance, Kanpur (Figure 3.2a) can be considered as a type A airport connected with three airports such as Ahmedabad, Delhi, and Mumbai. On the contrary, type B airports such as Dehradun (Figure 3.2b) which is connected to domestic (Jammu, Chandigarh, and Pantnagar) and international (Lucknow, Varanasi, Delhi, Mumbai, Ahmedabad, Hyderabad, Kolkata, and Bengaluru) airports and serves more than ten direct destinations.

3.3.2 International Airports

Over the past few decades, international travel in India has increased significantly. As a result, more than 30 airports in India cater to scheduled international flights. Major international airports in India are: Delhi, Mumbai, Chennai, Kolkata, and Bengaluru. Similarly, airports such as, Hyderabad, Ahmedabad, Kochi, Lucknow, Pune, and Goa also serve numerous international destinations. Daily direct flights operate from these

FIGURE 3.2 Instances of domestic airports' connectivity in India from type A (red) airport (a) and type B (yellow) airport (b).

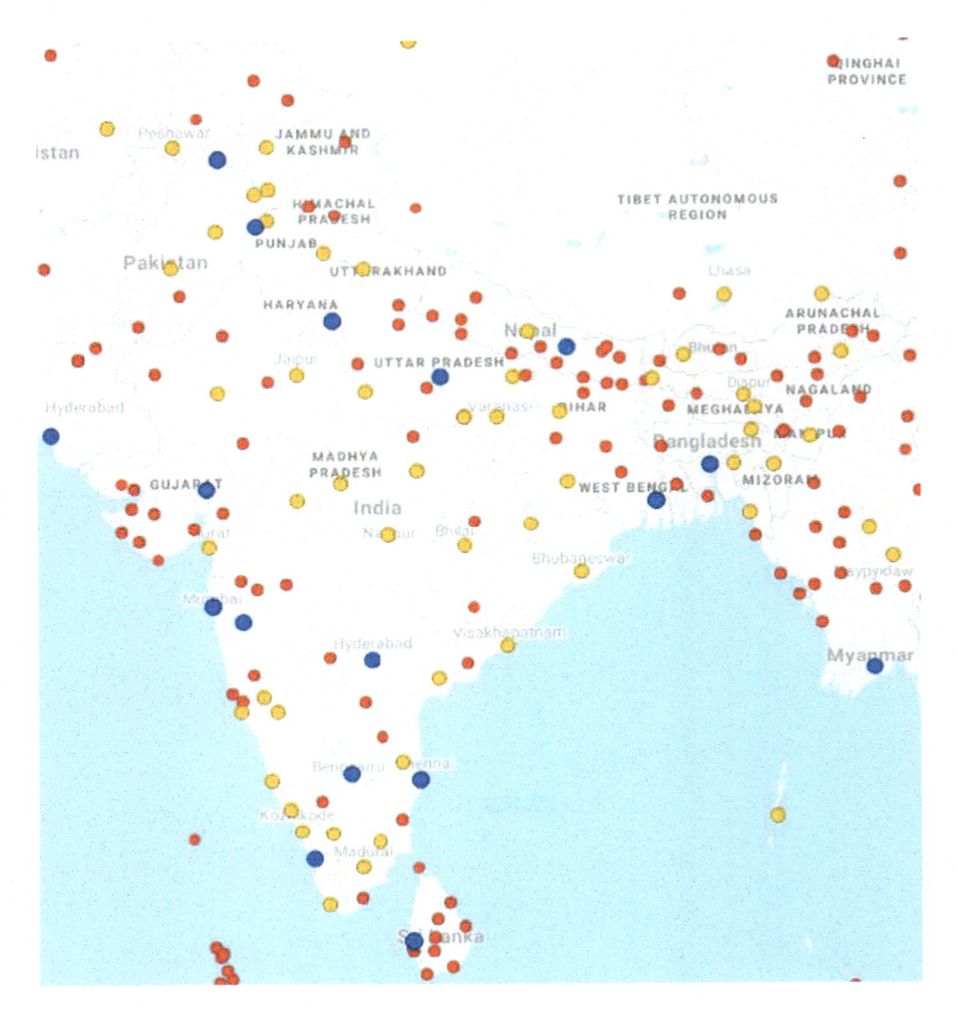

FIGURE 3.3 Map of major operational airports in India.

destinations to different parts of the world including Europe, North America, the Middle East, and Asia-Pacific.

Airports in India are catering to domestic and international flights operating in different IATA traffic conference areas. In terms of passenger traffic volumes, according to the Annual Airport Traffic Report (2019), the Delhi airport was ranked 17th (68,490,731) and the Mumbai airport was ranked 41st with 47,055,740 passengers worldwide. In Figure 3.3, major domestic and international airports operating in the civil aviation space in India are presented.

Table 3.2 presents airport codes using IATA and ICAO identifiers (Airport codes, n.d.) and the operations of the airport (domestic and/or international). The table also presents the number of direct destinations (Flight connections, n.d.) the airport is connected to using a color-coded scheme, i.e., red <5, yellow >10, and blue >50.

TABLE 3.2
Airport codes and their operational status

Airport			Operations		
City	IATA Code	ICAO Code	Domestic	International	Destinations
Adampur	AIP	VIAX	✓		🔴
Agartala	IXA	VEAT	✓		🔴
Agatti	AGX	VOAT	✓		🔴
Agra	AGR	VIAG	✓		🔴
Ahmedabad	AMD	VAAH	✓	✓	🔵
Allahabad	IXD	VIAL	✓		🔴
Amritsar	ATQ	VIAR	✓	✓	🟡
Asansol	RDP	VEDG	✓		🔴
Aurgangabad	IXU	VAAU	✓		🔴
Bagdogra	IXB	VEBD	✓	✓	🟡
Bathinda	BUP	VIBT	✓		🔴
Belgaum	IXG	VOBM	✓		🟡
Bengaluru	BLR	VOBL	✓	✓	🔵
Bhavnagar	BHU	VABV	✓		🔴
Bhopal	BHO	VABP	✓		🟡
Bhubaneswar	BBI	VEBS	✓	✓	🟡
Bhuj	BHJ	VABJ	✓		🔴
Bidar	IXX	VOBR	✓		🔴
Bikaner	BKB	VIBK	✓		🔴
Calicut	CCJ	VOCL	✓	✓	🟡
Chandigarh	IXC	VICG	✓	✓	🟡
Chennai	MAA	VOMM	✓	✓	🔵
Coimbatore	CJB	VOCB	✓	✓	🟡
Cuddapah	CDP	VOCP	✓		🔴
Darbhanga	DBR	VE89	✓		🔴
Dehradun	DED	VIDN	✓		🟡
Delhi	DEL	VIDP	✓	✓	🔵
Dibrugarh	DIB	VEMN	✓		🔴
Dimapur	DMU	VEMR	✓		🔴
Diu	DIU	VA1P	✓		🔴
Gaya	GAY	VEGY	✓		🔴
Goa	GOI	VOGO	✓	✓	🔵
Gorakhpur	GOP	VEGK	✓		🔴
Guwahati	GAU	VEGT	✓	✓	🟡
Gwalior	GWL	VIGR	✓		🔴
Hubli	HBX	VOHB	✓		🟡
Hyderabad	HYD	VOHS	✓	✓	🔵
Imphal	IMF	VEIM	✓	✓	🔴
Indore	IDR	VAID	✓	✓	🟡
Jabalpur	JLR	VAJB	✓		🔴
Jaipur	JAI	VIJP	✓	✓	🔵

(continued)

TABLE 3.2 (Continued)
Airport codes and their operational status

Airport			Operations		
City	IATA Code	ICAO Code	Domestic	International	Destinations
Jaisalmer	JSA	VIJR	✓		🔴
Jammu	IXJ	VIJU	✓		🟡
Jamnagar	JGA	VAJM	✓		🔴
Jharsuguda	JRG	VEJH	✓		🔴
Jodhpur	JDH	VIJO	✓		🟡
Jorhat	JRH	VEJT	✓		🔴
Kalaburgi	GBI	VOGB	✓		🔴
Kandla	IXY	VAKE	✓		🔴
Kangra	DHM	VIGG	✓		🔴
Kannur	CNN	VOKN	✓	✓	🟡
Kanpur	KNU	VICX	✓		🔴
Khajuraho	HJR	VAKJ	✓		🔴
Kishangarh	KQH	VIKG	✓		🔴
Kolhapur	KLH	VAKP	✓		🔴
Kolkata	CCU	VECC	✓	✓	🔵
Kochi	COK	VOCI	✓	✓	🔵
Kullu	KUU	VIBR	✓		🔴
Leh	IXL	VILH	✓		🟡
Lengpui	AJL	VELP	✓		🔴
Lilabari	IXI	VELR	✓		🔴
Lucknow	LKO	VILK	✓	✓	🟡
Ludhiana	LUH	VILD	✓		🔴
Madurai	IXM	VOMD	✓	✓	🟡
Mangalore	IXE	VOML	✓	✓	🟡
Mumbai	BOM	VABB	✓	✓	🔵
Mysore	MYQ	VOMY	✓		🔴
Nagpur	NAG	VANP	✓	✓	🟡
Nanded	NDC	VAND	✓		🔴
Nashik	ISK	VAOZ	✓		🔴
Pakyong	PYG	VEPY	✓		🔴
Pantnagar	PGH	VIPT	✓		🔴
Pasighat	IXT	VEPG	✓		🔴
Pathankot	IXP	VIPK	✓		🔴
Patna	PAT	VEPT	✓	✓	🟡
Porbander	PBD	VAPR	✓		🔴
Portblair	IXZ	VOPB	✓		🔴
Puducherry	PNY	VOPC	✓		🔴
Pune	PNQ	VAPO	✓	✓	🟡
Raipur	RPR	VERP	✓		🟡
Rajamundry	RJA	VORY	✓		🔴
Rajkot	RAJ	VARK	✓		🔴
Ranchi	IXR	VERC	✓		🟡

TABLE 3.2 (Continued)
Airport codes and their operational status

Airport			Operations		
City	IATA Code	ICAO Code	Domestic	International	Destinations
Salem	SXV	VOSM	✓		🔴
Shirdi	SAG	VASD	✓		🟡
Shillong	SHL	VEBI	✓		🔴
Shimla	SLV	VISM	✓		🔴
Silchar	IXS	VEKU	✓		🔴
Srinagar	SXR	VISR	✓	✓	🟡
Surat	STV	VASU	✓	✓	🟡
Tezpur	TEZ	VETZ	✓		🔴
Tiruchirapalli	TRZ	VOTR	✓	✓	🔴
Tirupati	TIR	VOTP	✓		🟡
Toranagallu	VDY	VOJV	✓		🔴
Thiruvananthapuram	TRV	VOTV	✓	✓	🟡
Tuticorin	TCR	VOTK	✓		🔴
Udaipur	UDR	VAUD	✓		🟡
Vadodara	BDQ	VABO	✓		🔴
Varanasi	VNS	VIBN	✓	✓	🟡
Vijayawada	VGA	VOBZ	✓		🟡
Visakhapatnam	VTZ	VOVZ	✓	✓	🟡

Based on Table 3.2, there are 58 type A, 33 type B, and 10 type C airports operating in India. Among these, most of the passengers pass through type A airports followed by type B, and type C. This leads to high congestion at type A airports which, at times, results in operational challenges to airline carriers, airport management, and inconvenience to passengers. Distribution of the passengers (domestic and international combined) among the major airports in India (Rathore et al., 2020) is presented in Figure 3.4.

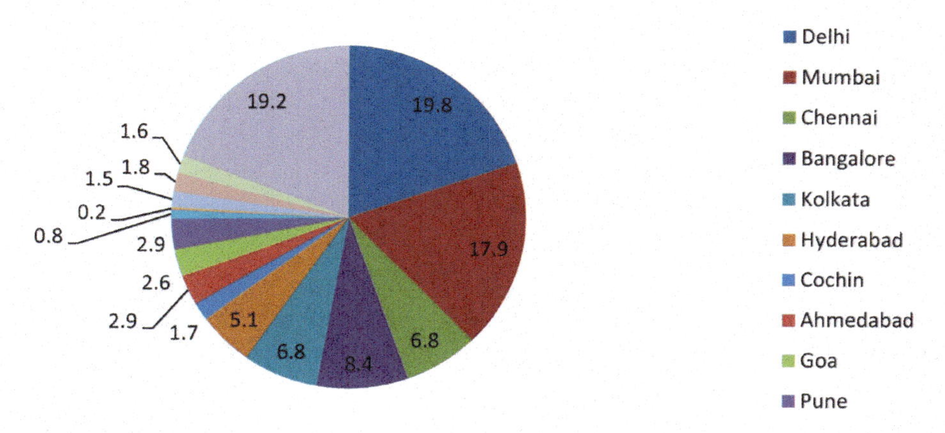

FIGURE 3.4 Passenger traffic distribution in 2018 among the major airports in India.

As evident from Figure 3.4, Delhi, Mumbai, and Bengaluru are the busiest airports in India catering to almost 60% of the passenger traffic. Also, airports such as Chennai, Kolkata, Pune, and Hyderabad cater to more than 25% of the passengers combined. The remaining passenger load is shared among the other airports spread across different regions of the country.

3.4 AIRLINE OPERATORS IN INDIA

3.4.1 THE EARLY DAYS

The first flight in India took place in 1911 by Henry Pequet (1888–1974) – a French pilot. This solo flight carried airmail and covered a short distance of 10 kilometers between Naini and Allahabad in northern India (Grant, 20017). In the early days of aviation in India, primarily mails were delivered between different cites namely Bombay, Karachi (now in Pakistan), and Madras. However, by 1925, airports were built in different cities of India to connect more cities by air transport. Civil aviation in India was established in 1927 and the first commercial flight was operated in 1929 from London to Karachi by Imperial Airways. As the aviation activity was developing in India, the need for professional flying was realized and as a result India's first flying school was established in Burma (now Myanmar) for commercial pilot training and licensing purposes.

The first India-based airline operator was the Tata Airlines owned by J.R.D. Tata. In 1932, the first flight was operated by Tata Airlines from Karachi to Bombay. In the same year, more flights were operated between different Indian cities by Tata Airlines which carried both airmail and passengers. In 1940, Hindustan Aeronautics Limited (HAL) was established in Bangalore for aircraft maintenance and other aircraft-related operations. In 1946, Tata Airways was branded as Air India and started operating internationally in 1948. The first international flight that flew from the Indian shores was from Bombay to London. Later, through the Air Corporations Act 1953, the government of India nationalized Air India. Similarly, the other domestic operators who were operating in different parts of India were merged into a single entity named as Indian Airlines. In 1960s, under the aegis of the government of India, Air India functioned with a larger fleet of aircraft catering to multiple locations in and outside India.

Similar to airport codes, airlines are also assigned a unique code. Among other aspects, airline codes help the Air Traffic Control in managing the flights by identifying the airline and the flight operating in their jurisdiction. IATA code for Air India is AI and ICAO code is AIC.

3.4.2 MAJOR DEVELOPMENTS

In 1990, the government of India made significant changes in the aviation sector and adopted the 'Open Skies' policy. Under this structure, both national and foreign airlines could operate their scheduled and non-scheduled cargo services from any airport in India. Furthermore, due to the change in the aviation policy, private

TABLE 3.3
Major India-based domestic and international airlines in 2018

Airline	Fleet Size	Daily Flights	Destinations: Domestic	Destinations: International	IATA Code	ICAO Code	Call Sign	Code Share Partners
Air India	221		57	45	AI	AIC	AIR INDIA	12
Indigo	285	1500+	63	24	6E	IGO	IFLY	14
Vistara	45	200+	40		UK	VTI	VISTARA	
Spice Jet	91		48	8	SG	SEJ	SPICEJET	
Air Asia	33		23		I5	IAD	RED KNIGHT	
Go Air		330	27	9	G8	GOW	GO AIR	

airline operators such as Sahara Airlines and Jet Airways also emerged on the scene in early 1990s. By 1995, more than 40 airlines were operational in India. In another significant development in 1995, International Airports Authority of India (IAAI) and National Airports Authority (NAA) were merged as AAI.

TABLE 3.4
Timeline of major events in the Indian aviation industry

Year	Event
1911	First flight carrying mail took place in Allahabad.
1927	Department of Civil Aviation was created.
1929	First international commercial flight between London and Karachi.
1932	First intercity flight between Karachi and Bombay.
1940	Hindustan Aircraft Ltd. was established.
1953	Nationalization of Aviation Industry. Air India came into existence.
1970	Deregulation Act Passed.
1972	International Airports Authority of India (IAAI) was formed.
1981	Vayudoot, a regional government-operated airline, was formed.
1986	National Airports Authority (NAA) was constituted.
1987	Bureau of Civil Aviation Security was established.
1992	De-regularization of Civil Aviation sector.
1994	Air Taxi Scheme was launched.
1995	Airports Authority of India (AAI) was formed by merging IAAI and NAA.
2007	Merger of Air India and Indian Airlines. National Aviation Company of India Ltd. Formed.
2016	National Civil Aviation Policy was introduced.

To boost the aviation sector, in 2004, the government of India relaxed the norms for the India-based airlines to operate on international sectors. In the changed policy, private airline operators who were operational with a fleet size of at least 20 aircraft for more than five years were allowed to operate internationally. That resulted in a merger of many regional airlines into a larger subsidiary. However, by 2019, many private airlines experienced a slowdown in the industry due to increasing operational costs, low profit margins, severe competition, rise in fuel costs, and so on. Due to this, airline operators such as Jet Airways and Jet lite, Air Costa ceased their operations in 2019. Despite this, in 2019, Air India was ranked 18th globally in carrying international passengers (197,464) and currently, Air India operates with a fleet of more than 120 aircraft serving around 90 domestic and international destinations (Air India, n.d.). To capture the gradual increase in the aviation operations and development in India, a summary of the major events is presented in Table 3.4.

3.4.3 Emergence of Low-Cost Carriers (LCC)

The LCCs or budget or no-frills airlines largely operate on short-haul sectors providing limited or no on-board services unless an extra charge is paid to avail catering and/or baggage services (Forsyth, 2010). The emergence of LCCs changed the landscape of the aviation industry in multiple dimensions. For instance, the competition between the airlines has increased manifolds triggering competitive pricing strategies and threatened the traditional business models (Michaels & Fletcher 2009; Belobaba, 2011). With the advent of the LCCs, the business models of the airline operators have changed regarding passenger seat allocation, catering policies, reservation options, baggage allowance, and so on. Low-cost airlines not only offer cheap fares but also help connect the destinations which otherwise were difficult to reach by air transport. Also, the increased frequency of LCC flights offer flexibility to the passengers giving them more options to fly to their desired destinations. For the aviation industry, LCCs proved to the boon emerging as the main contributor in the passenger traffic for the short haul operations (Forsyth, 2010). Also, intense competition between the airline carriers enforced tight utilization of their resources increasing the efficiency of aviation operations for airports and airline operators (Thirunavukkarasu & Nedunchezian, 2015).

Due to considerable changes in the aviation policies and growth driven by low-cost business model of the carriers, there has been a surge in the low-cost operators in the emerging markets such as India. In India, LCC operations have stimulated the demand for air travel significantly (Krämer et al., 2018; Wang et al., 2018) and dominate the aviation industry in the presence of traditional full-service carriers (Deeppa & Ganapathi, 2018). Furthermore, despite high taxes, heavy airport charges and increased competition, the success of LCCs in India is attributed to market demand and flexible connectivity between centers (Saranga & Nagpal 2016; Sakthidharan & Sivaraman, 2018). Overall, in the Indian aviation market, LCCs have played significant role in popularizing air travel for the masses. Based on the information provided on the official websites of major LLCs operating in India, IndiGo, Vistara, Go Air, Spice Jet, and Air Asia are operational in the domestic and international circuits (Table 3.3).

- IndiGo is a major LCC with a domestic market share of approximately 40% in December 2020. IndiGo is the largest airline operator in India (sixth in Asia) with respect to fleet size and number of passengers carried. In 2018–2019, the airline carried over 60 million passengers across 63 domestic and 24 international destinations. The IATA code for IndiGo is 6E, ICAO code is IGO and the call sign is IFLY. The carrier operates 1500 daily flights spanning across domestic and international sectors with a fleet of 285 aircraft (Indigo, n.d.).

- Vistara is a joint venture between Tata Sons and Singapore Airlines. The airline has a domestic market share marginally under 3%. The airline serves to 40 destinations with over 200 daily flights through a fleet of 45 aircraft. The IATA code of Vistara is UK, ICAO code is VTI and the call sign is VISTARA (Vistara, n.d.).
- The market share of GoAir is around 8%. The airline operates over 330 daily flights catering to 36 domestic and international destinations combined with a fleet of 57 aircraft. The IATA code of GoAir is G8, ICAO code is GOW and the call sign is GO AIR (Go Air, n.d.).
- Spice Jet is another major LCC with a market share of around 13%. The airline operates more than 600 daily flights with a fleet of 91 aircraft and serves 56 destinations. The IATA code of Spice Jet is SG, ICAO code is SEJ and the call sign is SPICEJET (Spice Jet, n.d.).
- Air Asia is a joint venture between Tata Sons and Air Asia Investments Ltd. The carrier operates with a fleet size of 29 aircraft to 21 destinations capturing 3% market share. The IATA code for Air Asia is I5, ICAO code is IAD and the call sign is RED KNIGHT.

A summary of operations of the major airlines operating in India is presented in Table 3.3.

Similarly, the market share of major airlines is presented in Figure 3.5. The operators share the passenger traffic through their domestic networks.

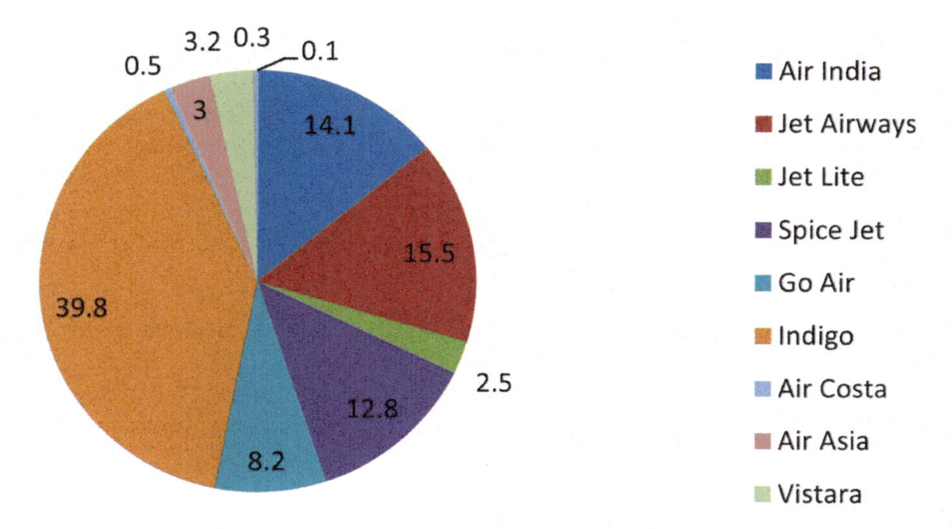

FIGURE 3.5 Market share in 2018 of domestic airline carriers in India.

The LCCs lack a specific or generic relationship model in dealing with airports, however they do demonstrate similarities in their business operations; but not all are equally successful in capturing the market (Forsyth, 2010). Regardless, there is strong commonality among the LCCs as they have agile mobility in operations and possess high flexibility to adjust their services depending on the requirements of the sectors they operate in (Forsyth, 2010). Unlike LCCs, Air India is the major carrier in India covering more than 100 domestic and international destinations combined. However, in recent times, Indigo has also emerged as a major airline operator with a fleet of 285 aircraft covering more than 1500 daily flights across their domestic and international networks. Other operators also function with a reasonable fleet size catering to a range of destinations in and outside India. However, the market share of these airlines in India (Rathore et al., 2020) varies significantly as presented in Figure 3.5.

3.5 CONCLUSION

In the last few decades, the aviation industry in India has witnessed policy and infrastructural changes leading to significant growth in the sector. For instance, the number of airports, airlines, passenger handling capacity, and operation of scheduled flights have increased manifold since 2000. With growing workforce and considerably large middle income group population, India is expected to be the largest civil aviation market by 2030. Due to the surge in demand for leisure and business and government support in developing the infrastructure, private sector participation is expected to be a dominant factor for investment opportunities in the Indian aviation sector. Regarding aircraft movement and passenger traffic in India, domestic travel accounts for more than three-fourths of the business. International operations account for more than half of the revenue generated from air freight traffic. Based on the passenger traffic, Delhi, Mumbai, Bengaluru, Kolkata, Hyderabad, and Chennai are the largest airports in India. The aviation industry is highly competitive among the LCCs which primarily operate in the short haul segment. Consistent increase in demand, aviation-conducive policies, and infrastructural development are the key drivers to shape the aviation sector in India in the next decade. Therefore, to achieve higher aircraft movement, growth in passenger traffic, rise in freight, and increase in airline operators, it is imperative to invest resources from government and private institutions.

CHAPTER QUESTIONS

Q1. Discuss the significance and scope of DGCA's operations and their impact on the civil aviation operations in India.

Q2. Discuss five differences among the different types of airports in India. How can the passenger traffic among the busy airports be distributed toward less busy airports?

Q3. Comment on type A, type B, and type C airports and list five domestic and international destinations for each airport type.

Q4. Discuss the impact of the government policy on the aviation industry in India. Identify two benefits and limitations of the policy from the perspective of airline operators.

Q5. Comment on the impact of LCCs in the aviation industry in India. Suggest two areas of improvement for LCC-conducive operations.

REFERENCES

Agrawal, A. (2020). Sustainability of airlines in India with COVID-19: Challenges ahead and possible way-outs. *Journal of Revenue and Pricing Management. 20*, 457–472.

Airport codes. (n.d.). Country codes India https://airportcodes.io/en/country/india/

Air India. (n.d.). About Air India www.airindia.in/about-airindia.htm

Annual Airport Traffic Report. (2019). Data and Statistics: Monthly Reports www.panynj.gov/airports/en/statistics-general-info.html

Belobaba, P. P. (2011). Did LCCs save airline revenue management? *Journal of Revenue and Pricing Management, 10*(1), 19–22.

Deeppa, K., & Ganapathi, R. (2018). Customers' loyalty towards low-cost airlines in India. *SCMS Journal of Indian Management, 15*(2), 42–48.

DGCA. (2021). Organizational structure www.dgca.gov.in/digigov-portal/?page=jsp/dgca/topHeader/aboutDGCA/organisation/org_Manual/Organisation%20Manual.pdf

DGCA. (2021). Regional Directorates www.dgca.gov.in/digigov-portal/?dynamicPage=RegionalOffices/0/50001 1/viewApplicationDtlsReqPDF

Flight Connections. (n.d.). India Airports www.flightconnections.com/ FlightConnections

Forsyth, P. (2010). Airport competition: The European experience. In P. Forsyth, D. Gillen, J. Muller & H-M. Niemeier (Eds), *Competition between Major and Secondary Airports: Implications for Pricing, Regulation and Welfare* (pp. 59–76), Routledge.

Go Air. (n.d.). About us www.goair.in/about-us

IATA. (2018). Potential and Challenges of Indian Aviation www.iata.org/en/pressroom/pr/2018-09-04-02/

IBEF. (2017). Indian Aviation Industry. www.ibef.org/download/Airports-February-2017.pdf

IBEF. (2021). Indian Aviation Industry www.ibef.org/industry/indian-aviation.aspx

Indigo. (n.d.). About us www.goindigo.in/about-us.html?linkNav=about-us_footer

Krämer, A., Friesen, M., & Shelton, T. (2018). Are airline passengers ready for personalized dynamic pricing? A study of German consumers. *Journal of Revenue and Pricing Management, 17*(2), 115–120.

Michaels, L., & Fletcher, S. (2009). Competing in an LCC world. *Journal of Revenue and Pricing Management, 8*(5), 410–423.

Rathore, H., Nandi, S., & Jakhar, S. K. (2020). The future of Indian aviation from the perspective of environment-centric regulations and policies. *IIMB Management Review, 32*(4), 434–447.

Sakthidharan, V., & Sivaraman, S. (2018). Impact of operating cost components on airline efficiency in India: A DEA approach. *Asia Pacific Management Review, 23*(4), 258–267.

Saranga, H., & Nagpal, R. (2016). Drivers of operational efficiency and its impact on market performance in the Indian Airline industry. *Journal of Air Transport Management, 53*, 165–176.

Spice Jet. (n.d.). About us www.spicejet.com/Fleet.aspx?web=1&wdLOR=c1861922A-4846-4C1D-8A0A-905C8D30A483

Thirunavukkarasu, A., & Nedunchezian, V. R. (2015). An analysis on domestic airlines capacity performance in India. *International Journal of Management, 6*(12), 38–49.

Vistara. (n.d.). About us www.airvistara.com/in/en/company-info

Wang, K., Zhang, A., & Zhang, Y. (2018). Key determinants of airline pricing and air travel demand in China and India: Policy, ownership, and LCC competition. *Transport Policy, 63*, 80–89.

4 Aviation Supply Chains

CHAPTER OBJECTIVES

At the end of this chapter, you will be able to

- Describe supply chains and its different operational phases.
- Distinguish between procurement, transformation, and distribution aspects of the chains.
- Understand an overarching view of aviation supply chains.
- Identify the key components of airline supply chain operations.
- Identify the major aspects related with airport supply chains.
- Understand key decision-making factors involved in the aviation supply chains.

4.1 INTRODUCTION

The focus of this chapter is supply chains, both in a general sense as well as from an aviation-centric point of view. Traditionally, in a supply chain the stakeholders, viz. the suppliers, manufacturers, distributors, and retailers combine their efforts and resources to create value for the consumer. Since supply chains are subject to the changing market needs they become difficult to manage over time. To counter such challenges, organizations implement policies at strategic, tactical, and operational levels. Within the purview of the aviation industry, decisions regarding the configuration of the supply chain, allocation of resources, and which processes each stage will perform, are critical to ensure the success of the chain operations. There are multiple separate supply chains operating concurrently in the aviation industry at a given time. For instance, the airline supply chain differs from the chain operations of an airport. Regardless, in each of these systems, value is created at different stages with a network of organizations working toward making a product or service conform to industry standards, which is crucial for aviation operations.

4.2 SUPPLY CHAIN MANAGEMENT

A supply chain is a coordinated system of organizations, information, people, and resources that transforms raw materials and components into finished products or

DOI: 10.1201/9780203731338-4

services. In a traditional supply chain, suppliers, manufacturers, distributors, and retailers integrate their efforts to create value for the consumer. This is achieved through a range of interconnected activities to meet customer demand as efficiently as possible. In general, supply chain is the management of operations to maximize customer value through sustainable competitive advantage achieved by responsiveness, low cost, or quality. In other words, a supply chain can be considered as a set of activities involved in meeting the customer's requirements through delivering goods and/or services (Chopra & Meindl, 2016, p. 14). In supply chains, all stakeholders serve as suppliers and customers at various business-to-business (B2B) and business-to-consumer (B2C) transactions. In this chapter, a customer (B2B) is distinguished with the consumer (B2C) or the end user for whom the value is generated through various supply chain stages.

The cumulative sum of all supply chain activities is to make outsourcing efficient, reduce inventory and transportation costs, meet competitive deadlines, reflect on the global nature of operations, and satisfy the ever-increasing demand for customization. To achieve this, a complex network of organizations shares tangible and intangible resources including information and strategies after establishing a significant degree of trust among the stakeholders. For effective supply chains, resource utilization is critical and can be streamlined by using appropriate quantitative and qualitative approaches at different levels in various supply chain stages. In general terms, the success of supply chain(s) is based on the efficient management of goods, services, processes, information, and capital.

Present-day organizations make consistent efforts to reduce expensiveness (cost and time) in the supply chain stages by slashing down unproductive activities through investing in research and development of new products and services. Continuous improvement in products, processes, and services is imperative for sustainable supply chains. This is achieved at all stages in the chain by using various statistical approaches and mathematical modeling techniques to keep check on product quality and process parameters. The objective is to make the products and services available at a competitive price by making optimal use of material, information, and financial flows. An overview of a generic supply chain is presented in Figure 4.1.

Supply chain operations begin downstream with the consumer demand. Based on the market survey response or the outcome of a feasibility study, the marketing department shares the findings and places request with the management to respond to the market needs. If approved, the finance department releases the required funds for the production department which assesses their operational capacity to meet the demand. At this stage, necessary equipment upgrade is considered along with material procurement requirements. Also, aspects surrounding licenses, copyrights, or intellectual property are addressed and permissions are obtained to ensure that all legalities are taken care of before the commencement of operations. If need be, the existing workforce is trained or skilled workforce is hired. Following this, transformation takes place to convert inputs such as raw materials into useful outputs which are distributed downstream through the network. Across the chain, processes are monitored and controlled through appropriate mechanisms by conducting quality checks, conforming to industry standards, and addressing the sustainability considerations. Throughout the

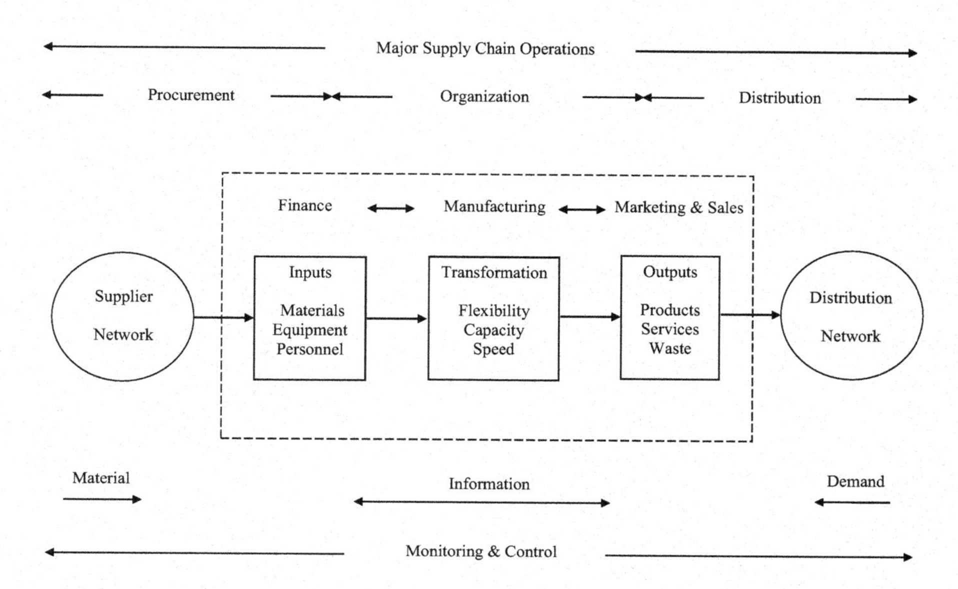

FIGURE 4.1 Overview of a supply chain.

chain operations, information is exchanged upstream and downstream. To achieve this, different strategies are devised across the supply chain as discussed in the following sections.

4.2.1 Strategies and Challenges

Supply chain operations or activities are broadly classified into three categories: strategic, tactical, and operational. Strategic activities are based on long-term goals ranging from few years up to few decades. It primarily considers supply chain design aspects such as determining the number and capacity of production facilities or location of the supply chain sourcing channels. In this step, considerations including forming long-term strategic alliances and make-or-buy decisions are the focus of the strategy. At this stage, strategy decides how the organization will compete in the industry. Business strategies are made by the top-level management for the entire organization and covering all the departments of the organization. Whereas, the planning horizons for tactical problems are of intermediate term which range from a couple of months to a year or are based on seasonal cycles. In this stage, departmental strategies are devised for each department by their heads. Each department such as production and marketing, establishes their strategies supporting the overall business strategy to achieve organizational objectives and goals. Inventory aspects, quality considerations and logistics operations, monitoring and control decisions form the core of the issues during this phase. Finally, operational aspects such as unscheduled maintenance and other unforeseen aspects are the major areas covered on the day or near to the day of operations.

Supply chains are dynamic and evolve over time catering to a range of market needs. Therefore, a lack of communication between supply chain partners, over-reliance on historical data than following the current trends, or not understanding supply chain partners' capabilities create challenges and may amplify the gaps between the expected and the actual demand upstream the chain. This phenomenon is known as the *Bullwhip* effect. Various supply chain stages have conflicting objectives based on their Key Performance Indicators (KPI) that makes supply chain operations challenging. Conflicts in supply chain management are not uncommon and at times unavoidable due to, but not limited to, the following factors.

- Complexities of culture.
- Global nature of operations.
- Cost-cutting initiatives.
- Un-updated information.
- Power control.

However, conflicts in supply chain management can be minimized, if not eliminated completely by using the following strategies:

- Effective communication up and down the stream in the chain.
- Information sharing among trusted partners.
- Aligning KPIs of the stakeholders.
- Regular inter-organizational meetings to resolve issues.
- Regular job rotation of the employees to gain multi-department awareness.

Despite best efforts, conflicts across the supply chain operations happen, especially in today's business practices which are agile and highly susceptible to uncertain environments. Contradictory objectives such as maximizing the market share, maintaining minimum production outputs, maximizing quality, minimizing noise levels, and so on are encountered in the chain operations on a regular basis. In real-world instances, it is not always possible to achieve every objective to the extent the decision-maker desires. Decision makers in manufacturing and service industries experience dilemma dealing with situations that have multiple objectives. With conflicting objectives, it is necessary to establish a hierarchy of importance among them so that lower-priority objectives are tackled only after the higher-priority objectives are satisfied. The trade-off with multiple objectives is achieved by reaching to a satisfactory level for each objective as opposed to achieving only a single objective.

4.2.2 Procurement

Procurement is a function of ordering, receiving, inspecting, and managing supplies. The objective of the procurement process is to ensure that materials procured conform to *quality* standards, in the appropriate *quantity*, from the right *sources*, at the best possible *price*, delivered at the right *place*, using suitable mode of *transportation*, at the scheduled *time*. Another major objective of the procurement process is to establish and maintain continuous flow of materials and minimize as many operational bottlenecks as possible to avoid disruption of subsequent processes. Materials vary in their utilization, and their significance is based on the degree of perceived value and risk, expensiveness (time and cost) and availability. Therefore, it is imperative to adopt different procurement strategies to suit the demands of the required supplies. To initiate the procurement process, management personnel identify key suppliers or supplier groups in line with the company's business objectives. This is to ensure that suppliers share common values, adopt ethical business practices, practice industry work ethics and are willing to share the information as needed. Suppliers can be classified in different categories such as Tier-I and Tier-II. This classification can be based on supplier–buyer relationship depending on the materials required. Generally, Tier-I suppliers provide components directly to the manufacturer whereas Tier-II suppliers are further upstream in the chain and provide materials to their Tier-I counterparts creating a network, as need be.

Close relationships are developed with both supplier categories based on the value they generate to the organization. Periodically, performance reports are generated for each supplier to measure their utility and their impact on the firm's business performance. The aim of this activity is to develop and maintain a seamless procurement process from the suppliers who generate profit to the organization and help create value for the consumer. In doing so, the following procurement steps are considered. In Figure 4.2, the procurement process involving a network of suppliers is presented.

In step one, the type of purchase is recognized and is categorized in three major aspects i.e., new, used, and modified. In the new purchase category, the material is no procured before and is a new entrant in the procurement list. Whereas, in the used category, the supplies procured earlier are ordered again without any changes in the specifications. However, in the modified category, content which has been procured earlier is ordered with updated or modified specifications of the required material.

In step two, specifications of the purchase is identified. In this step, conformance to industry standards and product requirements is essential. Also, at this step, patents

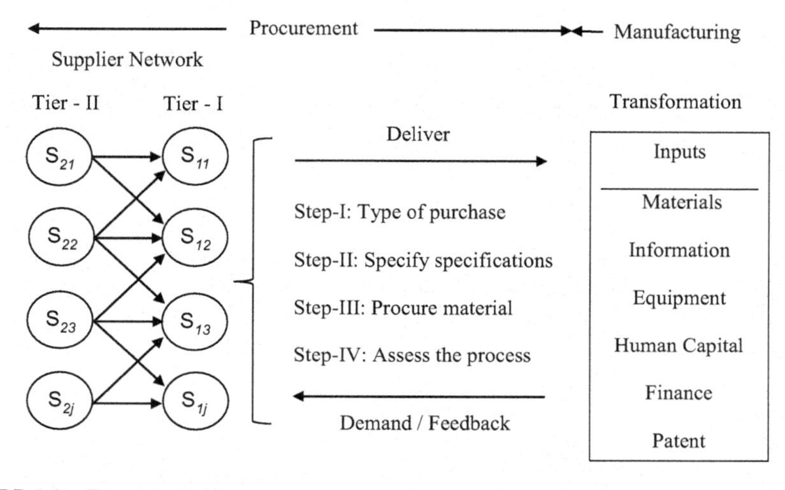

FIGURE 4.2 Procurement process.

or license-related issues are addressed to ensure ethical and legal aspects related to procurement and production. It is important to know that procurement of complex products require sophisticated (real-time data based) analysis prior to purchase.

In step three of the procurement process, a quote from the potential suppliers is generated and the order is placed after negotiations surrounding price, quantity, quality, and lead times. Once the order is received, an inspection of the material is carried out and records are updated to maintain the inventory levels. Also, payments are made as agreed between the involved stakeholders and materials are shipped for processing downstream.

Finally, in step four, the effectiveness of the procurement process is assessed to ensure that user requirements are satisfied. Through this, the value created by the procurement process is established by providing and receiving the feedback to and from the stakeholders involved in the procurement process.

In these steps, the major procurement concerns are the cost of materials, cost on materials, timely delivery of the materials to have smooth flows to avoid delays, and maintaining required inventory levels consistently. Once the materials are procured, quality assurance and pre-processing aspects are considered before the supplies are shipped for processing at the production facility. Throughout the supply chain activities, information technology and human resources departments play a critical role in ensuring smooth functionality in the inter- and intra- organizational processes.

4.2.3 Manufacturing

Transformation includes all activities necessary to implement and manage manufacturing operations, capturing flexibility in moving products into, through and out of the production facilities. This phase of operations considers the ability to make a wide range of products in a timely manner with optimal resource utilization. To achieve the desired level of manufacturing flexibility, planning, and execution must extend beyond the confines of the manufacturer to other stakeholders of the supply chain. Moreover, this enables management to coordinate the efficient flow of materials across the supply chain and reflect on the changing market needs through strategically planning the manufacturing operations. In this phase, among other input sources, supplies are procured primarily in the form of material and equipment. During transformation, production schedule, technology, and training supported by appropriate logistics play a critical role in achieving sustainable operations. Furthermore, compliance to industry standards enhances the quality of the final deliverable and helps reduce waste. During transformation, in each step, there are variables (*assignable* and *natural*) that have a significant impact on the processes (P_1 to P^n). Assignable variables correspond to random events (equipment breakdown, operator error) whereas natural variables capture the predicted scenarios which are inherent to the environment (infrastructure, old equipment). In Figure 4.3, an overview of the transformation phase is presented.

The primary objective of the production strategies is to plan and design the production schedule subject to budgetary limitations, scarcity of resources, time constraints, and mitigate the effect of variations in the processes. This can be achieved by

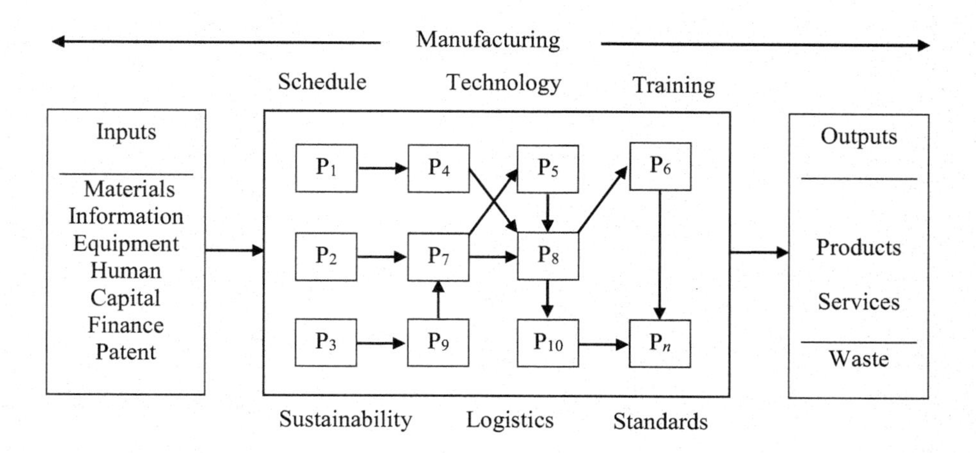

FIGURE 4.3 The manufacturing phase of a supply chain.

- Regularly monitoring the processes.
- Conforming to quality requirements.
- Troubleshooting and error handling.
- Optimally using resources.
- Integrating processes in line with organizational objectives.
- Discounting local objectives and focusing on organizational benefits.
- Increasing process flexibility and reducing variability.

In this phase of supply chain operations, raw material gets transformed and converted into final deliverables. At this stage, among other aspects, the consideration is to maintain the optimal inventory levels along with appropriate resource utilization to gain the economies of scale. Manufacturing processes also require equipment upgrades and constant product changes result in various investment scenarios leading to expansion opportunities. For this, Engineering Economy based studies help identify suitable returns on investment by comparing the competing.

Production strategies determine, by often incorporating various optimization tools and techniques, to find out; whether to manufacture the products in-house or to outsource it from other reliable supply sources. In such situations, organizations analyze their production capacity and the associated expenses and compare it with the outsourcing cost. If the outsourcing proposition is expensive than the total manufacturing cost, then it is recommended to manufacture the product in-house. However, to manufacture products in-house, few other aspects are considered such as whether skilled work force is available, are legalities taken care of, and whether the product can be made in a cost-effective manner. Moreover, it needs to be established through feasibility studies whether there is a long-term demand or high volume of product requirement to justify the investment for in-house production. Outsourcing is another option and is generally an attractive way out to satisfy the demand, especially when the organization deals with a low volume of products. Other major factors considered for outsourcing include; non-availability of skilled work force, better resource utilization elsewhere, or legalities that prevent the organization from manufacturing the products such as patents or copyright issues.

Once the decision to manufacture or outsourcing the products is made, the next significant consideration is to establish the optimal inventory levels to ensure smooth operations. Improper inventory management leads to expensiveness. Traditionally, capital is tied up in inventory through warehouse costs, insurance, breakage, or spoilage. However, there are other scenarios where inventory-driven costs are observed, such as; if unused, inventory loses value over time (*devaluation costs*), customers return products for a full return (*return costs*), product reaching their end-life (*obsolescence costs*), and utilization of money than investing in inventory (*opportunity costs*). In supply chain operations, quantity discounts or forward buying happens frequently and during this process, it is imperative to keep the total inventory cost as low as possible.

One of the primary objectives of the transformation phase is to plan the operations subject to the scarcity of resources such as material, capital, and time. Optimal resource utilization helps in sustainable operations. Market trends also dictate the type of the products to be manufactured; however, manufacturers are often inclined to produce most profit contributing products for their business. Optimization techniques such as linear and non-linear programming using advanced algorithms help address and solve production problems. During transformation phase, efficiency is critical. Production strategies need to be established so that a high degree of efficiency is achieved in the involved processes. Efficiency is the state at which a business is producing required quantities by utilizing the least possible number of resources. The objective is to achieve balance between resource utilization and production without compromising on the product quality. Time is a perishable resource but in general seldom gets priority over cost or quality. Equipment idle time, except maintenance, is expensive in all respects such as monetary aspects, productivity, customer satisfaction, employee morale, and soon. During transformation, the resource allocation problem arises because resources such as men and machines have varying degrees of efficiency to perform different activities or tasks under varying conditions. In this scenario, the objective is to assign several resources to an equal number of activities so as to minimize total cost/effort or maximize total profit/efficiency of the allocation.

Traditionally, manufacturing systems are broadly classified into continuous and intermittent processes based on the product variety and volume. More specifically, continuous processes follow the product layout, where a product flows between workstations at a nearly constant rate. This production process is suitable when high volumes of similar items are produced. For instance, in an aviation manufacturing plant where engines, wheels, and other parts are added to the fuselage as it moves along the assembly line. Whereas, in the intermittent processes, products of various designs move in different patterns between workstations to produce more than one product type.

Furthermore, production of merchandize using manpower, machines, tools, and chemical processing can be classified either based on the demand forecast (*make-to-stock*, MTS), customer requirements (*make-to-order,* MTO), or using the hybrid techniques (*make-to-assemble,* MTA). In MTS, the units are produced and stocked before the customer demand (*push system*). In push systems, process is initiated in anticipation of customer orders and the process is based on the speculation of demand as opposed to the actual demand. On the contrary in MTO, items are produced after customer orders are realized (*pull system*). In pull systems, operations are initiated in response to a customer order and this process is reactive in approach following the Just-in-Time (JIT) philosophy. It is important to know that most of the processes are a combination of push and pull approaches, such as MTA in which standard modules are produced in anticipation of the demand and assembled based on customer specifications. This creates a '*push-pull boundary*' in the supply chain. Supply chain processes are initiated depending on the timing of their execution relative to the customer demand and are separated from the pull–push boundary. To achieve this, integrated material flows are essential for smooth manufacturing operations. The integrated material flow system coordinates all storage and material handling elements and optimizes the material-in and -out flow. Furthermore, it synchronizes supply with demand and helps execute the strategic plan with minimal disruptions. In the manufacturing process, this can be achieved by seeking answers to the questions presented as follows:

- What is being done?
  - Is it a scheduled manufacturing process or a rework?
  - Is it an administrative task?
- When is it being done?
  - Is it carried out during the regular work hours?
  - Is it carried out during the extended shifts?
- Who is doing it?
  - Is it being carried out at the floor personnel level?
  - Are managerial level staff involved in the process?
- Where is it being done?
  - Is it carried out at the department, division, or process level?
  - Is it carried out at the job floor or elsewhere?
- How long does it take?
  - Can it be completed in batches, shifts, or days?
  - Is it carried out for the short term or long term?
- How is it being done?
  - Is it carried out on a regular basis?
  - Is it a one-off job?
- With what is it being done?
  - Are appropriate equipment being used?
  - Does real-time information exchange take place?
- How is the process evaluation measured?
  - Is it in monetary amounts?
  - Is it in productivity aspects?

A process needs improvement if it; is slow in responding to customers' needs, faces considerable number of quality issues or errors, is not cost effective, is not flexible enough to cater to variable demands, encounters frequent bottlenecks, involves disagreeable work, results in pollution, or does not add value towards the consumables.

4.2.4 Distribution

Across the supply chain activities, three fundamental flows occur along the operations: information flow, material flow, and monetary flow. Among these flows, the flow of information is initiated from the end-user (*downstream*) toward the supplier (*upstream*). On the contrary, the flow of material is initiated from upstream suppliers and flows downstream. During information exchange and material flow, financial transactions take place between supply chain partners resulting in monetary flows. It is important to realize that the end-user is the source of revenue generation and as a result, organizations co-ordinate their efforts to enhance value for the consumers. One of the major challenges for a supply chain is to distribute products or provide service to the consumers by creating the equilibrium between consumers' requirements with the capabilities of the supply chain. Supply chains achieve this by designing and creating networks (Figure 4.4) such that the firm satisfies the customer request while minimizing the shipment cost.

In supply chain operations, logistics play a critical role in planning, controlling, and smooth movement of goods from source to destination. The scope of logistics ranges from procurement through to the delivery of product or service to the end-user. Therefore, logistics activities are majorly classified into supply logistics, production logistics, and distribution logistics. During and post-production, multiple extensive quality control tests are conducted to ensure that whether the outputs conform to the desired product specifications. After this, the products are shipped, using suitable modes of transportation, through distributors which consist of wholesalers and/or retailers catering to different markets.

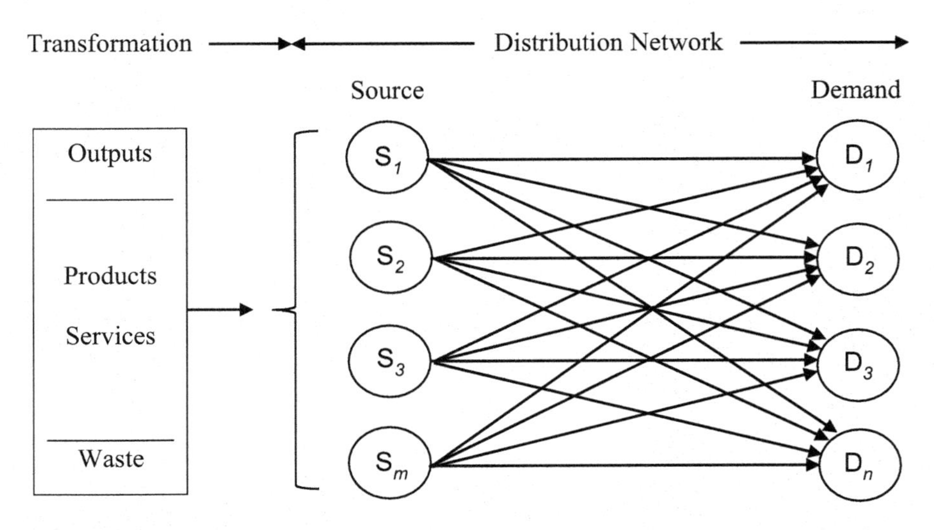

FIGURE 4.4 Distribution network in a supply chain.

4.3 AVIATION SUPPLY CHAINS

Like most other supply chains, an aviation supply chain is not a linear function of a set of activities catering to varied consumer demands. Similar to the conventional classification of supply chain activities mentioned in the earlier sections of this chapter, commercial aviation supply chains extensively rely on sourcing different goods and services from a range of business and government entities increasing the complexity of the aviation supply chain operations. However, it is pertinent to classify and view the aviation supply chain from the perspective of different stakeholders such as airports and airline carriers due to their significance in capturing and establishing value in the aviation operations. However, the cross-functional chain aspects such as service quality (Chapter 6) and sustainability (Chapter 9) are common to all aviation supply chains as presented in detail in these chapters.

In this chapter, the term 'supply chain' is used to denote the physical operations and distribution channels relating to the supply of products to end-users. Whereas, the term 'value chain' is referred to the processes and stages that add value to processes as they pass through the supply chain's various stages. In aviation supply chains, value is created by different stakeholders across different stages of the aviation operations and in each stage there is a network of organizations which generate value toward the end product or service. In the aviation industry, multiple supply chains are in action and work in tandem catering to the needs of a range of customers such as passengers, airlines, employees, concessionaires, tenants, and so on. Therefore, aviation supply chains are expensive, large in operations, dynamic, complex, and depend on multiple variables increasing the vulnerability of the chain operations. However, in this chapter, two major aviation supply chains are covered, i.e., airline chain and the chain related to airport operations.

4.3.1 AIRLINE SUPPLY CHAIN

Present-day airlines operate in an extremely competitive environment. Airline supply chains consist of suppliers ranging from regulatory bodies, equipment manufacturers, fuel providers, and catering services. Also, there are sequential airline operations which include flight schedule construction, aircraft fleet management, and crew scheduling using tight use of the airline's resources leading to complexity in the operations. In case of schedule disruptions, on the day of operations, the affected airline resources such as aircraft and crews, if need be, are allotted revised tasks to ensure that broken passenger itineraries are repaired. Crews such as the captain and the first officer (technical crew) along with the purser and airhostesses (cabin crew) are an important input resource to an airline. The utilization of these resources is subject to aircraft fleet and size, destinations covered in the flight network, and the flight frequency (Budd et al., 2020). Figure 4.5 presents an overview of a generic airline supply chain capturing different suppliers.

In the following section an insight into airline suppliers such as equipment manufacturer, fuel providers, and catering services is presented. Details on regulatory bodies is covered in this chapter. However, details on airline operations such as flight scheduling, fleet management, crew scheduling, and disruption management are covered in Chapters 6, 7, and 8.

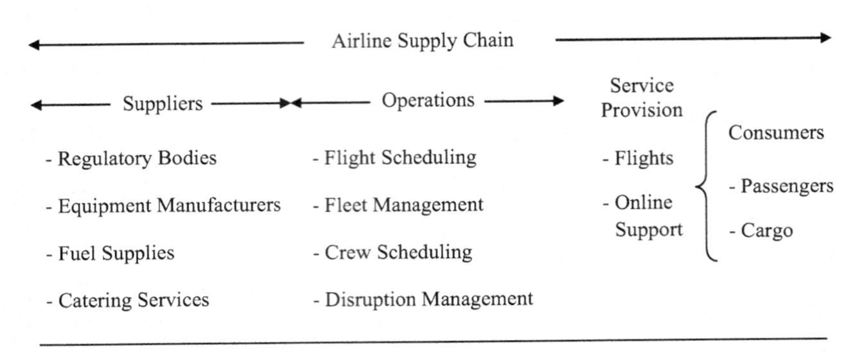

FIGURE 4.5 Overview of the airline supply chain.

4.3.1.1 Equipment Manufacturers

The aircraft is one of the most expensive resources to an airline. Airbus and Boeing are the two largest manufacturers supplying aircraft to the majority of the airlines worldwide. Airbus is a major equipment manufacturer based in Europe and has production facilities in France, Germany, and Spain. The company deals with both civilian and military equipment. Airbus aircraft operates with 2 to 4 engines with a seating capacity ranging from 100 to more than 500 passengers. Some of the most successful Airbus aircraft are A300, A320, and A380 used for short haul to long haul flights (Airbus, n.d.).

Boeing is the pioneer in civil aviation industry. The company is based in the United States and deals with both commercial and defense aircraft equipment. The seating capacity of Boeing's twin engine 737, 767, 787, 777, and 747 aircraft range from 80 to more than 550 passengers (Boeing, n.d.). Boeing's Dreamliner aircraft has created opportunities for the company through fuel-efficient and technologically advanced equipment making the company a leader in the market segment. Whereas, another aircraft manufacturer Bombardier is based in Canada and, similar to Airbus and Boeing, deals with both civil and military aircraft (Bombardier, n.d.). However, as compared to the other two aircraft manufacturing giants, Bombardier has a relatively smaller market share and the operations are limited in terms of the aircraft range.

In the aviation industry, almost all airlines operate through different fleets of aircraft. A fleet is a collection of aircraft of the same type such as Airbus 320 or Boeing 767. Within a fleet, all aircraft are identified with a unique alpha-numeric identifier known as a *tail number*. For equipment manufacturers, airlines place orders well in advance based on their business strategy.

Companies such as Rolls-Royce, Honeywell, and GE Aviation are among the major suppliers upstream the chain, supplying engines (ignition and/or turbine) and other aircraft parts to airlines worldwide.

Aircraft manufacturing operations are sophisticated and require high degree of precision in the equipment design. Research and development are an integral part of the aircraft manufacturing process and used extensively to mitigate process variations through different experimental design techniques. Experimental design is a technique in which a set of experiments are conducted by changing the input variables of a manufacturing process to observe corresponding changes in the response variable. This technique is useful to develop new products, improve existing product designs, or to achieve robustness in the design process. Robustness in the process can be achieved by making the process insensitive to uncontrollable factors or external sources of variations. More specifically, experimental design methodology emphasizes on designing the quality into the products and processes which is different from the conventional quality assurance practices such as reliance on product inspection (Roy, 2001) which is significantly critical in aircraft equipment design. During different phases of the aircraft manufacturing operations, experimental designs are carried out to ensure consistency and reliability in the final deliverable. Also, pull systems using the JIT approach are used along with push systems based on supplier capacities and manufacturing requirements. Following this, shipments are delivered to the airlines.

Based on the customers' requirements, aircraft manufacturers make changes, if needed, in the interior design of the equipment such as designing the overhead bins and customizing the seating arrangements. Most of the aircraft are equipped with over-water-flying capability, making them suitable for the intercontinental operations. Airlines which operate over the water bodies need such equipment in their fleet of aircraft to conform to industry standards. Modern-day aircraft function using real-time data management system during flight operations. This is especially critical in ensuring information exchange between the equipment manufacturer and the airline to deal with emergency situations.

4.3.1.2 Fuel Supplies

Aircraft fuel is produced in refineries after following a complex procedure of turning raw materials into different grades such as Avgas 80, Avgas100, Avgas100 low lead, and Jet A1 fuel (Breidenthal, 2019). Among the fuel grades, Jet A1 fuel is the premium quality product and is the mostly used fuel type in aviation industry. The fuel is provided by the suppliers and stored in large tanks on or near the airport from where it is shipped to the aircraft for refueling. An aircraft is a massive equipment that carries passengers, luggage and cargo items. In an aircraft, generally the fuel tanks are situated in the wings and at the center of the fuselage. Regarding fuel consumption requirements for a flight, many aspects are considered before deciding the optimum fuel quantity such as the flight duration, total carrying weight, aircraft type, the weather conditions, contingencies, and industry regulations.

4.3.1.3 Catering Services

Catering is an integral and important part of the airline industry. Airlines outsource or have their own catering services. Major airlines have established their own catering services (at hubs) to serve hundreds of flights on a daily basis and cater to different dietary preferences of the passengers. In serving different regions of the world, catering services procure a variety of produce from across the world to make a range of meal types daily. The catering facility works continuously, ensuring fresh supplies are procured and processed for consumption. Stocks are placed and monitored through technically advanced mechanisms and, as need be, placed under temperature-controlled environments. Stringent quality checks are performed during and after the entire process of preparing millions of meals annually. At the food dispatch facility, once the meal is prepared, cutlery and other condiments are included in the trays which are placed, along with the beverages, on different trolleys for the first, business, and economy class passengers. After this, food trolleys are shipped to the airport prior to the departure and loaded onto the aircraft for in-flight services.

4.3.1.4 Agents

In the aviation industry, traditionally, travel agents act as a retailer for the consumer. Alternatively, travel agents can be viewed as an interface between the service provider, i.e., the airline and the passenger. Travel agents are required to get accredited from IATA to sell tickets for both domestic and international travel and earn commission from the service provider for selling tickets on their behalf. However, with the increased digitization of many aspects of the aviation industry, ticket booking through various online options is fast replacing the traditional business practices. In the present-day scenario, consumers not only do business directly with airlines but also compare ticket prices with highly competitive Internet-based portals before selecting the best offer from the number of available options. However, through self e-purchase, at times, many subtle policy nuances are overlooked by the consumers which otherwise can easily be discussed with the travel agents to avoid any inconvenience during the travel. For instance, with the advent of budget airlines or LCCs and the mushrooming of regional carriers, the aviation business model has changed considerably. This includes highlighting the base fare and excluding various indispensable taxes in the published ticket price, no inflight entertainment, on-demand meal purchase, no baggage allowance, and so on. Some airlines charge the air fare from passengers based on their body weight (CBS News, 2013). Notwithstanding, the role of a travel agent is considerably significant in modern-day travel in providing the best possible, cost effective, and convenient itinerary to the consumer from a plethora of Internet-based booking options. Furthermore, based on the historical data and seasonal patterns, the travel agents are likely to be aware in advance of the special offers and travel deals being offered by the airline. They also have a fair insight of the travel insurance policy of the carrier, changing rules, and regulations on the on-arrival VISA policy of the visiting destination.

For airlines, the supply sources and operational aspects can also be captured as the internal and external factors which are significant in establishing airline supply chain network. The relative influence of these internal and external factors in airline supply chain is presented in Figure 4.6.

Based on the internal and external factors, the strategic decisions of an airline include the following aspects.

- Identify airport locations and their operational capacity.
- Establish information sharing mechanisms among the partner airlines.
- Build and maintain stakeholder partnerships.
- Outsource required sources of supply.

Airline operations are seldom limited to a specific geographic location, but rather stretch to multiple regions the airline serves. To enhance the reach of their operations,

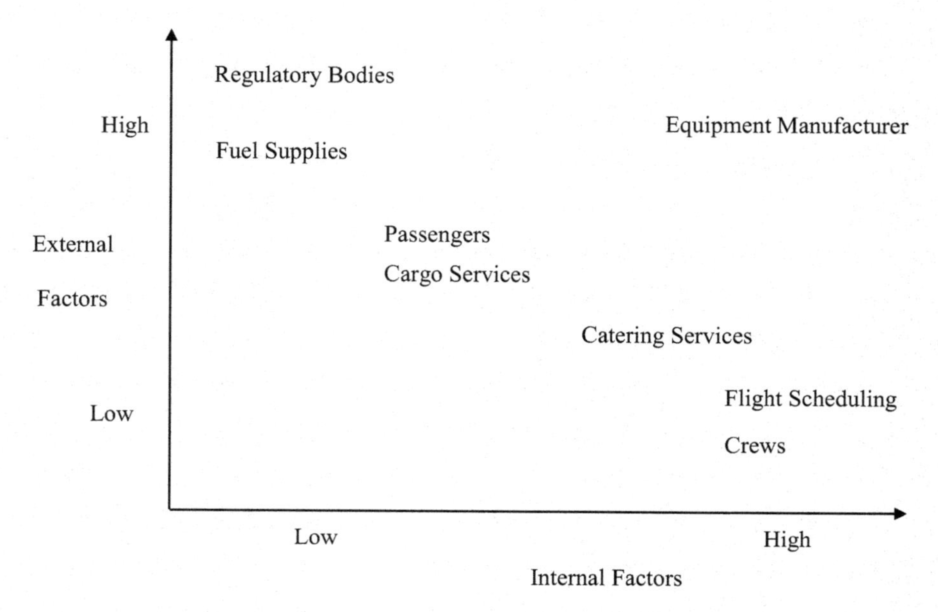

FIGURE 4.6 Influence of factors in airline supply chain.

airlines establish partnerships with other airline carriers and share their resources to enhance their network. Through this, the airlines expand their supply chains and provide consistent goods and services to a wider consumer base, i.e., to passengers and freight operators (Palmer, 2020). To achieve this, airlines create network and dyadic relationships with other service providers by strategic alliances through coopetition in some markets (*code sharing*) and competing in the other segments. Airlines view code-sharing operations as an extension of their own. Through this, streamlined check-ins, by sharing information, significantly improve the passenger experience in modern-day air travel. Airlines integrate their resources for

- Faster and flexible response to customer demands.
- Sharing information to run operations smoothly.
- Mitigating redundancy in effort, information, and planning.
- Improving efficiency and service.
- Minimizing uncertainty, errors, and delays.
- Eliminating non-value adding activities in the operations.

Similar to safety, trust is paramount in the aviation industry, especially within the airline operations. Trust is developed when an airline's performance history and the reliability of its supply chain linkages can be demonstrated through reliability, responsiveness, resilience, and relationships which are significant factors leading to satisfactory customer experience. Among others, key dimensions of trust are: reliability and competence. Reliability depends on prior experiences and in aviation supply chains, consistency and predictability over extended periods are likely to lead to reliability among consumers. Reliability is also based on the integrity of other stakeholders. A lack of congruence between claims and action may deteriorate the degree of reliability between airline operators and consumers.

4.3.2 Airport Supply Chain

Airports serve as the point of connection between airlines and passengers by providing facilities required for airline operations and passenger travel. Aviation supply chains are increasingly complex with airports as the focal entity of the aviation operations where the movements of airline carriers, passengers, and freight operators converge at different airport facilities. Another challenging aspect in the airport supply chains involve the coordination of various government institutions and private business entities to ensure policy implementation across the airport operations (Button, 2020). Airports add value in the passenger experience by providing complementary services and follow through with the airline to ensure a comfortable and enjoyable travel experience for their customers. During aviation operations, ground staff play critical role in ensuring smooth, safe, and efficient functioning of the processes both for the airline and the consumers. Figure 4.7 presents an overview of the airport supply chain capturing suppliers and a range of operations to ensure efficient functioning of airport's facilities.

Majorly, functioning of airport operations depends on the regulations advised by international and national regulatory bodies, government mandates, and airline operators. Airports are also responsible to carry out a range of operations such as providing logistics for air traffic control, gate allocation to arriving flights, baggage handling services in collaboration with airlines, managing real-time flight information, immigration services, and security and maintenance to passengers and cargo operators. Details on aviation regulatory bodies have been covered in Chapter 2. However, detailed insights on airport operations such as Air Traffic Control, gate allocation, baggage handling, immigration services, and other aspects are covered in Chapter 5.

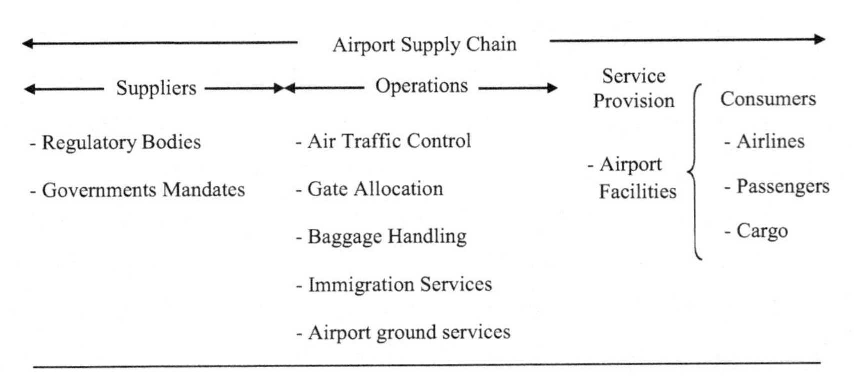

FIGURE 4.7 Overview of airport supply chain.

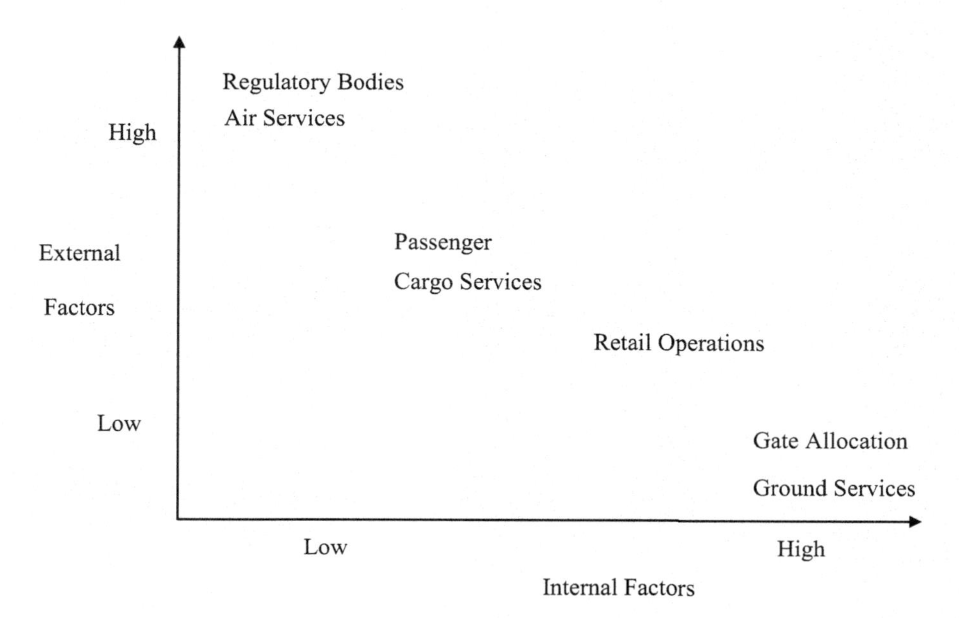

FIGURE 4.8 Influence of internal factors in airport supply chain.

For airports, the supply sources and operational aspects can also be captured as the internal and external factors which are significant in establishing an airport's supply chain network. Figure 4.8 presents the relative influence of these internal and external factors in airport supply chain.

Similar to airlines, for airports, various internal and external factors are crucial in establishing a robust supply chain. Based on the internal and external factors, the strategic decisions for an airport include the following aspects;

- Gate allocation and ground operations' capacity.
- Establish information-sharing mechanisms with the airlines and industry regulators.
- Build and maintain partnerships with retail outlets and cargo operators.
- Outsource sources of fuel supply.

Strategic planning is crucial for both airline and airport supply chain operations. Decisions about the configuration of the supply chain, allocation of resources, and what processes each stage will perform are critical in ensuring the success of aviation operations. For instance, airlines and airports continually identify the opportunities to minimize costs upstream and downstream the chain. Aviation supply chain design decisions are long-term and expensive to reverse, i.e., such decisions must consider uncertainties including market trends, consumer behavior for passengers, airlines and cargo operators, and industry regulations.

4.4 CONCLUSION

In the present day, aviation supply chains operate in a highly competitive environment. They function as a sophisticated and well-coordinated system comprised of international regulatory bodies, expansive set of facilities, and expensive resources, highlighting information sharing among the aviation industry stakeholders. Within the aviation industry, the supply chains are determined by four significant stakeholders at different levels, namely; regulatory bodies, airports, airlines, and consumers. Different strategies are implemented within these systems through numerous sequential and concurrent activities to meet user demand as efficiently as possible. In airline supply chains, the suppliers include regulatory bodies, equipment manufacturers, fuel providers, and catering services. Meanwhile, the airport supply chains are responsible for providing logistics for air traffic control, gate allocation for arriving flights, baggage handling services in collaboration with airlines, managing real-time flight information, immigration services, security, and maintenance for passengers and cargo operators. While the aviation supply chains may be highly dynamic, their design decisions are expensive to reverse in spite of the uncertainties involved in their operations.

CHAPTER QUESTIONS

Q1. Discuss the scope and impact of aviation supply chains on international travel. Identify five areas in aviation supply chains for improvement.
Q2. Identify five key aspects of an airline supply chain and discuss its two benefits and limitations for the passengers.
Q3. Evaluate the role of internal and external factors in the airline supply chain and discuss the impact they have on the decision making on airline operations.
Q4. Identify five key aspects for the supply chain of an airport and discuss its two benefits and limitations for the passengers.
Q5. Evaluate the role of internal and external factors in the supply chain of an airport and discuss the impact they have on the decision making on airport operations.
Q6. Identify five similarities and differences between the supply chain of an airline and the supply chain operations of an airport.
Q7. How can airline supply chain and airport supply chain operations be streamlined for efficient ground operations? Suggest five points each for both supply chains.

REFERENCES

Airbus. (n.d.). Passenger aircraft www.airbus.com/en/products-services/commercial-aircraft/passenger-aircraft

Boeing. (n.d.). Products and services www.boeing.com/commercial/

Bombardier. (n.d.). Aircraft https://businessaircraft.bombardier.com/en/aircraft

Breindenthal, R. (2019). Environmental concerns in general aviation. In E.A. Hoppe (Ed.), *Ethical Issues in Aviation* (pp. 225–232). Routledge.

Budd, T., Intini, M., & Volta, N. (2020). Environmentally sustainable air transport: A focus on airline productivity. In T. Walker, A.S. Bergantino, N.S. Much, & L. Loiacono (Eds.), *Sustainable Aviation: Greening the Flight Path* (pp. 55–77). Palgrave Macmillan.

Button, K. (2020). Boulding, Brundtland, economics, and efforts to integrate air transportation policies into sustainable development. In T. Walker, A.S. Bergantino, N.S. Much, & L. Loiacono (Eds), *Sustainable Aviation: Greening the Flight Path* (pp. 29–54). Palgrave Macmillan.

CBS News. (2013). Plane travel by the Pound? Samoa Air says charging passenger's by weight is paying off. www.cbsnews.com/news/plane-travel-by-the-pound-samoa-air-says-charging-passengers-by-weight-is-paying-off/

Chopra, S., & Meindl, P. (2016). *Supply Chain Management: Strategy, Planning, and Operation (6th ed.)*. Pearson.

Palmer, W. (2020). Sustaining flight: Comprehension, assessment, and certification of sustainability in Aviation. In T. Walker, A.S. Bergantino, N.S. Much, & L. Loiacono (Eds.), *Sustainable Aviation: Greening the Flight Path* (pp. 6–28). Palgrave Macmillan.

Roy, R. (2001). *Design of Experiments Using the Taguchi Approach: 16 Steps to Product and Process Improvement*. Wiley.

5 Airport Operations

CHAPTER OBJECTIVES

At the end of this chapter, you will be able to

- Identify different types of airport operations.
- Distinguish between apron, taxiway, and runway operations.
- Know the Air Traffic Control and baggage handling operations.
- Get an overview of airport competition.
- Know the significance of airport location and service quality.
- Understand passengers' consideration for airport selection.

5.1 INTRODUCTION

Airports are the major actors in aviation value chains, playing a significant role in aviation operations and serving as the point of entry/exit for millions of travelers around the world each year. They are the focal point of interaction for stakeholders associated both, directly (airlines, passengers) and indirectly (neighborhood communities) with the aviation industry. However, airports are significant not only for air transportation but also in the economic development of a nation through establishing trading and travel routes for passengers and freight (Bunchongchit & Wattanacharoensil, 2021).

On one side of the spectrum of aviation operations, the airlines manage their day-to-day flight operations using numerous airport facilities to serve passengers in a timely manner, whereas on the other side, various logistics-based aspects are managed by airport authorities to ensure smooth functioning of airport operations. In other words, almost all key actors associated with the aviation industry frequently communicate with one another to run a range of airport operations. Airports are dynamic facilities that require state-of-the-art infrastructure and technical expertise to perform complex functions and provide essential services, all in conformance with industry standards and regulations. They also contribute to the social and economic requirements of the stakeholders associated with the aviation industry within the region in which they operate (Baxter & Sommerville, 2011; Bilotkach et al., 2012; Florida et al., 2015). In this chapter, the focus will be on the operational aspects of different airport facilities from the perspective of airlines, airport operators, and passengers.

DOI: 10.1201/9780203731338-5

5.2 OPERATIONAL ASPECTS

Construction of airports involves high fixed costs, whereas they function with comparatively low operating costs (Button, 2010, p. 59). Airports provide aviation infrastructure to airlines (runways, aprons, and taxiways), freight operators (cargo terminals, ground handling, and warehouses), and facilities to passengers such as baggage handling, internal transportation, and lounge services (Bilotkach & Mueller, 2012; Suárez-Alemán & Jiménez, 2016). Furthermore, other service providers such as retail outlets, fuel suppliers, surface transport companies also align their business objectives with the scale and scope of airport operations. Airports attract not only the airlines but also the freight operators to use their facilities for their business operations. Freight companies, especially the air cargo operators, working in tandem with trucking businesses have significantly expanded the scale of freight operations. Due to this, flexible and competitive options are available for the freight companies to tap into the range of services available across the aviation operations. In the following sections, major airport facilities and their associated operations catering to commercial aviation practices are elaborated.

5.2.1 Airside Operations

Broadly, airport activities are categorized into airside and landside (or ground) operations. Landside operations include airport ground access areas such as the airport terminal building, the cargo facility, and the parking areas (de Barros, 2015; Janić, 2015). Whereas airside operations are related to the movement of aircraft which require large areas of land for air traffic control (ATC) operations (de Barros, 2015) and includes airport maneuvering areas such as the apron area, taxiway, and runway in the airport premises (Wilke et al., 2014; Ortega & Manana, 2016). An overarching view of different airport operations and aircraft flight stages is presented in Figure 5.1.

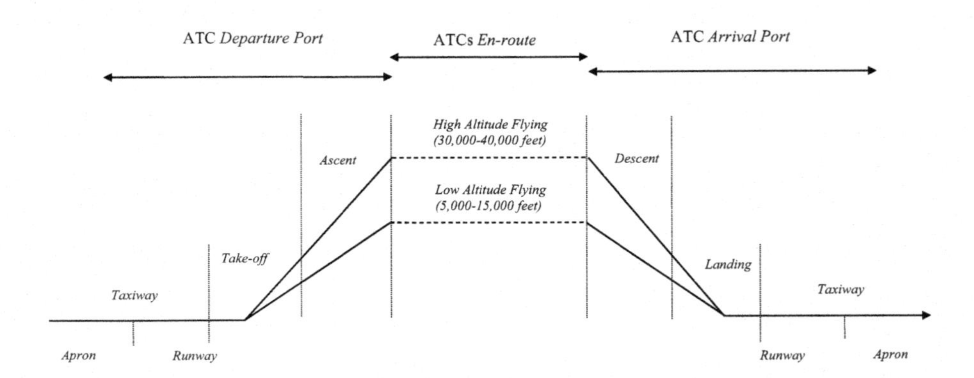

FIGURE 5.1 Airport operations and aircraft flight stages and aircraft flight stages.

5.2.1.1 Apron

The apron is an area at the airport where aircraft parking slots are situated. Aircraft are parked at the apron to let passengers board/alight the aircraft at the assigned gate of the airport terminal. For both airlines and airports, efficient operations at the gate are critical to achieve optimal resource allocation through minimum-cost operations. Therefore, it is imperative for the airports to manage the terminal operations through optimal gate allocation and alleviate aircraft congestion during departure/arrival of the aircraft. It is important to know that due to the surge of low-cost carriers (LCCs), airports provide customized facilities at specific airport terminals for the LCC operators (Forsyth et al., 2010).

At the apron, other aircraft specific ground services are also provided such as loading/unloading of baggage, cleaning, and refurbishing of the supplies for the next flight, aircraft fueling, visual aircraft maintenance checks, and so on. Apart from the aircraft, considerable movement of vehicles and personnel takes place at the apron. For instance, vehicles at the apron such as tugs to tow aircraft, snow removal devices during winters, and maintenance vehicles operate on the ground to provide necessary support and services to the arriving and departing aircraft (Wilke et al., 2014).

5.2.1.2 Taxiway

The taxiway connects the apron with the runway. After fulfilling all pre-departure flight checks, the aircraft leaves the apron and taxis to the runway where it is queued (if need be) before becoming airborne (Wilke et al., 2014). All departing aircraft are assigned a scheduled time to taxi, also called pushback time. For departing flights, one of the major objectives for the aircraft is to taxi and reach the runway within the specified time window for the scheduled take-off. Therefore, from an operational perspective, it is critical to coordinate aircraft turnaround operations to minimize delays and congestion at the taxiway through the management of landside and airside resources. For this, airport operations depend on the interactions between airport authorities, ground handling staff (loading/fueling), pilot, and the ATC as appropriate to facilitate safe, efficient, and scheduled movement of the aircraft. After ATC clearance at the port of departure, the aircraft gets on the runway before getting airborne to climb to the required flying altitude. To avoid turbulence and air traffic, mostly, the commercial jets fly at a higher altitude (30,000 to 40,000 feet) whereas smaller planes with unpressurized cabins fly around 10,000 feet and rarely go below 5,000 or beyond 15,000 feet.

Upon approaching the destination port, the aircraft leaves its flight level, descends and lands on the runway designated by the ATC at the port of arrival. After landing, the aircraft is directed from the runway to taxi and reach as early as possible to its allocated apron areas connected with vacant gates and park. However, at busy airports or hubs, gate assignment is a significant problem leading to delays in the ground operations. Among other aspects, airside operations at the airport include addressing scheduling problems. For instance, directing the aircraft to their allocated spaces in a timely manner to mitigate delay in airline operations and waste in airport resources.

5.2.1.3 Runway

The runway is an important area of the airport used only for aircraft take-offs and landings. Runways are located away from other airport services and facilities and specially designed to mitigate the adverse effect of winds during take-offs and landings. At any of the flight stages, while the aircraft is airborne, it may experience headwind (adds to aircraft speed), tailwind (negates aircraft speed), or crosswind (lateral push to the aircraft) impacting its functionality. Even though take-offs and landings are performed in small time windows, it is imperative to have conducive weather conditions to execute these critical flights tasks. Ideally, take-offs and landings are preferred with headwinds and without tailwinds or crosswinds. However, at the time of take-offs and landings, rarely does the weather stay calm, i.e., no headwind, tailwind, or crosswind. Therefore, depending on the orientation of the runway, the direction of the wind determines the direction of the aircraft take-off and landing.

As compared to a take-off, the landing maneuver is more challenging and relatively difficult to master. For a smooth landing, appropriate aircraft speed and angle is required at the right place for a touchdown on the runway. Similarly, for a take-off, an important aspect is the time duration between two takeoffs. It is critical to know that depending on the type and size of the aircraft and based on the weather conditions, aircraft generate engine power to get the required drag and lift during take-off. Due to this, an aircraft's take-off leaves a jet-stream of varying intensities and scale (wake vortex) which at times impacts the subsequent flight's take-off. Therefore, to maximize the utility of the runway and to avoid interference with the jet-stream trail from the take-off of the previous aircraft, time and/or distance-based separation constraints between take-offs are imposed (Janić, 2015). In other words, the separation constraint is based on the weight of the aircraft, ensuring that after the departure of the heavier aircraft, larger separation is required for the departure of a lighter one. Another separation constraint captures aircraft departures which follow similar departure routes (en-route separations). After ensuring the separation constraints from the previous flight of the aircraft, the next aircraft takes-off from the runway and reaches its intended flight level based on the information gathered from the ATC operators.

5.2.1.4 ATC Operations

The critical components of the aviation industry include complex and interdependent operations comprising of airports, airlines, and air navigation aspects (Button, 2010, p. 59). The ATC functions as a bridge between the airport and an airline's operations to ensure safe and efficient air traffic operations conforming to industry rules and regulations. The ATC operates for a designated airspace known as the 'jurisdiction of the ATC' and uses state-of-the-art technology and equipment, such as radio navigational aids and satellites to communicate with the airside and ground operations using real-time data exchange (Janić, 2015). The major objectives of ATC operations are to establish a safe distance (> 10 min flying time at the same level) or altitude (>1,000 ft) between flights when they are airborne, prevent aircraft collision with an on-ground obstacle, and maintain the flow of air traffic (Wilke et al., 2014). For this, ATC follows time-based and/or distance-based separation rules subject to the departure/arrival of an aircraft on the runway (Janić, 2015). During the entire flight operations, i.e., ground movement of the aircraft and while the flight is airborne, the location and trajectory information of the aircraft is shared across relevant stakeholders by the ATC. Through this, ATC ensures that communication between the ground resources and airspace management is established and maintained to ensure flight safety and to utilize airport capacity to achieve scheduled aviation operations.

5.2.2 Landside Operations

In the aviation industry, the role and significance of airports, airlines, and the ATC is critical. In this, airport and ATC signifies the infrastructural capability in dealing with the on-ground and air traffic which is dependent on the flight schedules generated by the airlines (Janić, 2014). Even though airlines have little or no control over airport surface transportation but airline operators, at times, also compete regarding the accessibility and availability of surface transport in their operational network. For safe, efficient and, convenient flight operations, airports streamline dynamic land and air operations and provide various services to airlines and passengers (Akyuz et al., 2019). Airlines and airports work in tandem to ensure a safe and convenient travel experience to the passengers through customer-friendly and efficient ground operations and services. For this, real-time flight information is shared between airports and airlines through display screens, airport websites and phone apps based on the real-time data collected from multiple sources to keep the passengers updated about flight departure and arrival status. Despite this, airport and airline operations face challenges in the form of technical issues, emergency situations, and regarding security regulations in the operations.

Among other functions, baggage handling and reclaim is one of the most important aspects in ground airport operations. Similarly, reclaiming baggage at the airport carousel is an important element in the travel experience for a passenger. Due to considerable increase in the passenger volume and movement of baggage across various airport channels, the likelihood of delays in baggage reclaims and losing or misplacing a baggage is increased at the arrival and/or departure airport (Davies, 2015; Koenig et al., 2019). However, the realization of not being able to claim the baggage is observed on the reclaim carousel at the port of destination (Davies, 2015). Regarding baggage delays, it is important to know that due to technological advancements in the immigration and custom services such as introduction of biometric passports and automated customs and passport control mechanisms, passenger throughput time is reduced considerably. This leads to earlier passenger arrival at the baggage reclaim carousels resulting in long wait times for passengers to claim their baggage (Graham, 2018). Nevertheless, delays in the baggage reclaim process and especially misplacement of the baggage leads not only to operational issues at the airports but also considerably affects an airline's service quality and reputation (Koenig et al., 2019). This leads to customer dissatisfaction and at times results in lost baggage claims which range between $200 and $3500 in compensation from the airline (Anand & Rajaram, 2016).

IATA also mandates its member airlines to keep track of baggage at check-in, aircraft loading, during transfers in between and on arrival to ensure safe return of baggage to passengers. Airlines and airports have also introduced several bag-locator mechanisms to track the location of checked-in baggage using online tracking through radio frequency identification technology, via phone apps to identify and locate the baggage on a real-time basis (Kang, 2016), and to minimize baggage switching instances (Anand & Rajaram, 2016). Therefore, for passengers, airports aim to provide safe travel with least possible through time (immigration services) and smooth baggage handling experience.

5.3 AIRPORT COMPETITION

In many instances, airports are owned by the governments which have control over major airport operations. However, airport ownership has varied over time and now they are publicly owned, privately owned or partially privately owned (Forsyth et a.l, 2010; Suárez-Alemán & Jiménez, 2016). Regardless of the ownership type, airports outsource a range of services from different private businesses (Button, 2010, pp. 63–64). Airports use marketing strategies to attract airlines and assist them in distinguishing their services through promotions, incentives, discounts on landing fee, and other pricing strategies including gate allocation and flight departure times (Forsyth et al., 2010; Morrell, 2010, pp. 11–26).

5.3.1 Airport Monopoly

With the change in regulations in the aviation industry from government-owned to privately owned airport facilities, there is an apprehension among the stakeholders about the monopolistic behavior (Morrell, 2010, pp. 11–26). The airport industry is highly competitive and seeks business propositions and opportunities to enhance profitability. Among airports, competition is primarily based on time, price, and connectivity (land and air) associated with the travel. Airport operations have low marginal costs, and any unused airport capacity is expensive for the facility. Therefore, competition is severe among the airports especially between those which are part of a dense airport network such as the airports located in Europe.

5.3.2 Airport Capacity

Airports are growing consistently to reflect on the increased demand and the trends and forecasts in the aviation industry. Therefore, airports aim to expand their operations and increase their capacity despite facing constraints such as fluctuation in lead times, logistical challenges, increasing costs, and environmental issues related with the expansion alternatives (Lutte & Bartle, 2017). To meet the demand, creating airport capacity is considered as one of the significant challenges for airport authorities to ensure smooth global aviation operations (Upham et al., 2012, p. 218). Generally, major hub airports are required to address the capacity issues often leading to pressure on their operations due to the structure and function of hub-and-spoke system (Lutte & Bartle, 2017). Especially, at hub airports congestion of traffic affects both landside and airside operations impacting arrival (gate allocation) and departure (aircraft separation) operations. To address the capacity issues, airports direct their efforts on the efficient use of both landside and airside resources to tackle the operational challenges regarding surface traffic, ecological effects, airport heating, and cooling systems and increased aircraft noise (Budd et al., 2013; Lutte & Bartle, 2017). Therefore, in airport operations economies of scale and scope are critical aspects which are linked to infrastructure, air traffic congestion, and passenger demand (Button, 2010, p. 65).

5.3.3 Airport Location

Another critical aspect in increasing the airport capacity is the airport location. Due to urbanization of cities, in many instances the expanse of cities has gone beyond the city bounds and has engulfed the airport which otherwise used to be on the outskirts of the city limits (de Barros, 2015). To build an airport in or around a city which already has an existing airport is considerably challenging compared to a city that has no airport. Existing airports have a monopoly in providing aviation services to airlines which include relationship-based pricing, incentivized gate allocation charges, and so on. Moreover, airports plan their operations surrounding the restrictions (environmental aspects) they operate under to enhance the opportunities (market share) they capture which is difficult for a new entrant to compete with (Forsyth et al., 2010). On the contrary, in case of more than one airport in the city or if the airports are located nearby, the competition among the airports challenges the monopolistic functioning, and the facility offering lower costs to customers benefits and enhances their market share (Forsyth et al., 2010).

Generally, airports have no or little impact in establishing a hub for major airlines. Therefore, large or major airports which are generally located within a city compete for the hubs for full-service carriers or as bases for LCCs by offering them subsidies and incentives (Forsyth et al., 2010). Whereas a secondary or smaller airport may function as a regional airport which are mostly established by converting the former military airbases for commercial flight operations (Forsyth et al., 2010; Button, 2010, p. 68). Such airports are often located away from the large airport or located in the vicinity of a major city and primarily compete to capture LCC traffic which is not committed to any specific airport. Public transportation such as monorail, high speed rail, or metro rail systems are convenient options connecting the secondary airports with city centers directly.

For LCCs, large and congested airports are not as attractive for operations as compared to the smaller airports (Button, 2010, p. 66). LCCs prefer this operational strategy to maximize their resource (aircraft, crews) utilization (Warnock-Smith & Potter, 2005). Through this approach, LCCs minimize the underutilization (aircraft and crew idle time) of their resources by avoiding congestion related with ground operations, turnaround times, gate allocation, and ATC clearance which is commonly encountered at the larger airports (Button, 2010, p. 68). Due to these reasons, the commercial utility of secondary or smaller airports has increased manifolds, especially in Europe and the United States which has many such facilities and serve numerous destinations with considerably short flight durations increasing the number of in-bound and out-bound flights at the secondary airports (Button, 2010, pp. 68–69). Therefore, the competition between major and secondary airports is increasingly becoming evident to the extent that in some instances secondary airports have significantly impacted the flow of traffic at the major airports (Button, 2010, pp. 59–76).

The airport network in Europe and the United States (especially in the northeast region) is highly developed and serves to a dense population resulting in high competition among the airports. However, in certain instances the competition between the airports is not direct since closely placed airports do not necessarily provide similar services and only compete for a small portion of the traffic. In few instances, the airports are preferred equally by the customers due to their geographical location (equidistant, convenience) or because the difference between the total cost associated with the travel is negligible between the competing airports. For instance, in South-East Asia, airports such as Bangkok (IATA; BKK), Singapore (IATA; SIN), and Kuala Lumpur (IATA; KUL) compete with one another and with the airports in China such as Guangzhou (IATA; CAN) for hub traffic due to their geographical proximity. Similarly, airports in the middle east such as Abu Dhabi (IATA; AUH), Dubai (IATA; DXB), and Doha (IATA; DIA) compete to attract airlines to base their airport facilities as for their hub operations. On the contrary, sparsely located airports do not have such a competition due to large distances.

5.3.4 Airport Service Quality

For passengers, while considering an airport for travel, one of the decision-making aspects is the airport service quality (Fodness & Murray, 2007). Even though the domain of service quality in the airport operations spans across a multitude of operations and lacks standardization, nevertheless, industry operators, researchers, and academics measure service quality through establishing and monitoring performance measures using a range of internal (service times, lost baggage) and external measures such as delays and connectivity issues (Yeh & Kuo, 2003). The internal service performance measures are helpful for benchmarking purposes whereas the external measures provide information about the operational attributes related to customer expectations (Fodness & Murray, 2007). Overall, airport service quality aspects are primarily addressed from the perspective of aviation industry stakeholders such as airport authorities, airline operators, and regulatory bodies. Airport service quality is also measured from the perspective of passenger convenience, information visibility, security, and immigration processing times (Fodness & Murray, 2007). However, the lack of inputs from the passengers is an issue that needs to be considered to further improve airport service quality (Correia et al., 2007; Fodness & Murray, 2007). Therefore, international aviation regulatory bodies such as Federal Aviation Administration and Airport Council International (ACI) have conducted numerous studies to capture different service aspects and measure the level of service for passengers, which is an evolving and significant issue in airport operations due to the pressing constraints airports function under (Correia et al., 2007).

At times, airport service quality is an important factor between competing airports to attract customers through the provision of tangible (infrastructure) and intangible (information) services (Fodness & Murray, 2007). For on-time flight operations, it is important that service quality of the landside operations addresses the changing needs of the aviation industry, such as reliability (flight arrival/departure times), increasing passenger volume, and safety and security in the aviation operations (Suárez-Alemán & Jiménez, 2016). Responding to safety and security requirements, airport authorities such as immigration services coordinate efforts and resources and exchange information with border control agencies at security checkpoints.

At the airports, the objective of airlines is to make the passenger transition as swift and smooth as possible. On the contrary, for airports the slow movement of passengers opens a window of opportunity to generate business for a range of retail outlets (Button, 2010, pp. 59–76). To make an airport attractive for passengers, airport marketeers engage in identifying customer needs from non-aviation revenue generation sources such as retail outlets and restaurants (Harrison, 1996). Therefore, based on airport customer review ratings, airports are reviewed and rated by organizations such as Skytrax and ACI on different operational aspects surrounding the terminal building (seating, cleanliness, signage), layout design (passenger flow, queueing time), and retail facilities (shopping, food, and beverage) (Bunchongchit & Wattanacharoensil, 2021).

5.3.5 Passenger Considerations

For air travelers, the airport is an intermediary or a facility that helps them connect with the destination(s) they are visiting for work or leisure. Therefore, air travelers choose the airport for their travel based on considerations such as available airport services or infrastructure of the airport, time of the travel, connectivity with other mode(s) of transportation, and the total cost of the travel. For instance, in Europe, especially the UK and Germany have a dense airport network along with significantly developed rail and motor way systems. Similarly, there is a competition between rail services and airports in some areas of the United States (especially northeastern region) and Japan. For traveling, speed, convenience, and safety is paramount, therefore other modes of transportation also provide travel alternatives which compete with the airports. However, the competition between airports and other intermodal options depends on numerous aspects such as the trade-off between total cost, connectivity options, convenience, and the travel duration (Button, 2010, pp. 65–67).

In considering the secondary airports (preferred by LCCs) as the point of origin or destination, at times, the travel cost to the airport and other overheads (parking, premium on services) associated with the secondary airports, which are generally located in the outskirts of the city, may exceed the cost of the air travel itself (Forsyth et al., 2010). Furthermore, in terms of service and accessibility, secondary airports are not as convenient as the major airports but attract LCCs and cost-sensitive passengers. However, for price-conscious travelers there is a downside of using budget airlines or LCCs who do not ensure connectivity with other flight operators or with other modes of transportation. On the contrary, surface transport operators adjust their operations to suit the arrival–departure times of the LCC carriers. Whereas airlines that operate on a large scale and have a well-established flight network (including alliances such as One world, Star Alliance, and SkyTeam) ensure passenger transfer in their itinerary

(Button, 2010, p. 70). In many instances, if needed, these operators delay the flights (within permissible time windows) to ensure that passenger itinerary is not broken.

5.4 CONCLUSION

Airports function as a connecting point for a range of stakeholders, attracting both the airlines and freight operators to use their facilities for business operations. Additionally, service providers such as multinational retail outlets, fuel suppliers, surface transport companies etc. also align their business objectives with airport operations. The airport industry is extremely competitive and hinges on time, price, and connectivity (land and air). Hence, safety and efficiency operations in the airport, whether airside or landside, are key areas of focus for authorities and airport regulatory bodies globally (White, 2012). However, for the air traveler there are a different set of considerations, such as the availability of airport services or infrastructure, time of travel to and from the airport, connectivity with other mode(s) of transportation, and the total cost of travel, based on which they decide to avail the services of one of the most important actors in the aviation value chain – the airport.

CHAPTER QUESTIONS

Q1. Identify three major airside operations at an airport. For each operation, discuss their role and significance in airport operations.

Q2. Identify three major landside operations at an airport. For each operation, discuss their role and significance in airport operations.

Q3. How does airport competition affect aviation operations? Identify three reasons and discuss their impact in the aviation industry.

Q4. How does Air Traffic Control function? Discuss five most significant operations of the ATC in aviation operations.

Q5. How does airport competition affect the passengers? Identify three reasons and discuss their impact on passenger travel choices.

REFERENCES

Akyuz, M. K., Altuntas, O., Sogut, M. Z., & Karakoc, T. H. (2019). Energy management at the airports. In T. H. Karakoç, C. O. Colpan, O. Altuntas & Y. Sohret (Eds), *Sustainable Aviation* (pp. 9–36). Springer.

Anand, S., & Rajaram, R. (2016). Ergonomics in airport baggage reclaim. *Indian Journal of Science and Technology*, *9*(11), 1–5.

Baxter, G., & Sommerville, I. (2011). Socio-technical systems: From design methods to systems engineering. *Interacting with Computers*, *23*(1), 4–17.

Bilotkach, V., Clougherty, J., Mueller, J., & Zhang, A. (2012). Regulation, privatization, and airport charges: Panel data evidence from European airports. *Journal of Regular Economics*, *42*(1), 73–94.

Bilotkach, V., & Mueller, J. (2012). Supply side substitutability and potential market power of airports: Case of Amsterdam Schiphol. *Utilities Policy*, *23*, 5–12.

Budd, L., Griggs, S., & Howarth, D. (2013). *Sustainable Aviation Futures (Transport and Sustainability),* pp. 3–35. Emerald.

Bunchongchit, K., & Wattanacharoensil, W. (2021). Data analytics of Skytrax's airport review and ratings: Views of airport quality by passengers types, *Research in Transportation Business & Management*, *41*, 100688.

Button, K. (2010). Airport competition: The European experience. In P. Forsyth, D. Gillen, J. Müller, H-M. Niemeier (Eds), *Countervailing Power to Airport Monopolies*. pp. 59–75.

Correia, A. R., Wirasinghe, S. C., & de Barros, A. G. (2007). Overall level of service measures for airport passenger terminals. *Transportation Research Part A*, *42*, 330–346.

Davies, H. (2015). Airports Commission: Final report. https://assets.publishing.service.gov.uk/government/uploads/system/uploads/attachment_data/file/440316/airports-commission-final-report.pdf

de Barros, A. G. (2015). Airport Planning and Design (1st ed.), Handbook of Transportation, Routledge.

Fodness, D., & Murray, B. (2007). Passengers' expectations of airport service quality. *Journal of Services Marketing*, *21*(7), 492–506.

Forsyth, P., Gillen, D., Müller, J. & Niemeier, H-M. (2010). Airport competition: The European experience. In P. Forsyth, D. Gillen, J. Müller & H-M. Niemeier (Eds), *Introduction and Overview* (pp. 1–10), Routledge.

Florida, R., Mellander, C., & Holgersson, T., (2015). Up in the air: The role of airports for regional economic development. *The Annals of Regional Science*, *54* (1), 197–214.

Graham, A. (2018). Airport economics and performance benchmarking. In *Managing Airports: An International Perspective* (5th ed.), pp. 68–95.

Graves, D. (1998). *UK Air Traffic Control: A Layman's Guide*, 3rd ed., Airlife Publishing: Shrewsbury.

Harrison, J. (1996). Airport retailing: A view of the future. *Airport Forum*, *26(2)*, 41–44.

Janić, M. (2014). Modelling the effects of different air traffic control (ATC) operational procedures, separation rules, and service priority disciplines on runway landing capacity. *Journal of Advanced Transportation, 48, 556–574.*

Janić, M. (2015). *The Air Traffic Control System*. The Handbook of Transportation, Routledge.

Kang, A. (2016). Delta introduces innovative baggage tracking process. https://news.delta.com/delta-introduces-innovative-baggage-tracking-process

Koenig, F., Found, P., & Kumar, M. (2019). Improving maintenance quality in airport baggage handling operations. *Total Quality Management*, *30*(1), 35–52. Lutte, R. K. & Bartle, J. R. (2017). Sustainability in the air: The modernization of international air navigation. *Public Works Management & Policy*, *22*(4), 322–334.

Morrell, P. (2010). Airport competition: The European experience. In P. Forsyth, D. Gillen, J. Müller & H-M. Niemeier (Eds), *Airport Competition and Network Access: A European Perspective* (pp. 11–26).

Ortega, A. S., & Manana, M. (2016). Energy research in airports: A review. *Energies* *9*(5), 349.

Suárez-Alemán, A., & Luis Jiménez, J. (2016). Quality assessment of airport performance from the passengers' perspective. *Research in Transportation Business & Management*, *20*, 13–19.

Upham, P., Maughan, J., Raper, D., & Thomas, C. (2012). *Towards Sustainable Aviation*. Routledge.

Warnock-Smith, D., & Potter, A. (2005). An exploratory study into airport choice factors for European low-cost airlines. *Journal of Air Transport Management*, *11*, 388–392.

White, J. R. (2012). FAA runway safety initiatives. www.icao.int/APAC/Meetings/2012_APRAST1/Bangkok%20RASG%20condensed.pdf

Wilke, S., Majumdar, A., & Ochieng, W. Y. (2014). Airport surface operations: A holistic framework for operations modeling and risk management. *Safety Science*, *63*, 18–33.

Yeh, C-H. & Kuo, Y-L. (2003). Evaluating passenger services of Asia-Pacific international airports. *Transportation Research Part E*, *39*(1), 35–48.

6 Airline Operations

CHAPTER OBJECTIVES

At the end of this chapter, you will be able to

- Understand the sequential airline schedule construction approach.
- Know about fleet assignment and aircraft routing aspects.
- Understand crew scheduling, crew pairing, and crew rostering.
- Comprehend sequential airline scheduling process through a problem instance.
- Know about integrated and robust airline scheduling approaches.

6.1 INTRODUCTION

Since the Wright brothers' first successful manned flight at the turn of the 20th century (Padfield & Lawrence, 2003), the aviation industry has witnessed unprecedented technological advancements in the past 120 years. Today, commercial airlines operate within a dynamic (ever-changing, fast-moving) and uncertain (variable demands, security issues) landscape. In a highly competitive industry such as this, whether an airline flourishes or tanks is determined by efficient planning and resource management at various stages of operations. And so, Operations Management (OM) methodologies such as lean philosophy and just-in-time have significantly reshaped operational aspects. Similarly, Operations Research (OR) models and techniques have greatly impacted planning and managing operations within the industry since the 1950s (Barnhart & Talluri, 1997). Furthermore, exponential growth in computing capabilities, both in hardware and software, and considerable advances in optimization algorithms have greatly aided in solving complex problems pertaining to airline scheduling. The successful implementation of optimization models in various airline operations led to the establishment of OR departments in many airlines, saving them millions of dollars annually. In the following sections, various airline schedule construction techniques such as the traditional sequential scheduling process, along with robust and integrated scheduling, are presented.

DOI: 10.1201/9780203731338-6

6.2 AIRLINE INDUSTRY

6.2.1 Airline Schedule Construction

Traditionally, airline scheduling follows a sequential procedure of flight schedule construction followed by solving fleet assignment, aircraft routing, crew pairing, and rostering problems. These problems are considered as planning problems and are broadly classified into strategy planning and operational planning. Applications of OM and OR approaches and techniques in airline operations range in both strategic and operational aspects in which strategic planning problems are solved for longer time horizon than operational planning problems. An illustration of these planning problems against a timeline is presented in Figure 6.1.

Long-term planning consists of flight schedule construction, fleet assignment, aircraft routing, and maintenance while short-term planning includes crew scheduling, revenue management, gate assignment, and dealing with irregular or disrupted operations. It is important to know that throughout the planning process it is assumed that an airline schedule would be implemented as planned without taking unforeseen events such as disruptions into consideration. Based on Figure 6.1, the traditional sequential approach of airline schedule construction is elaborated in the following sections.

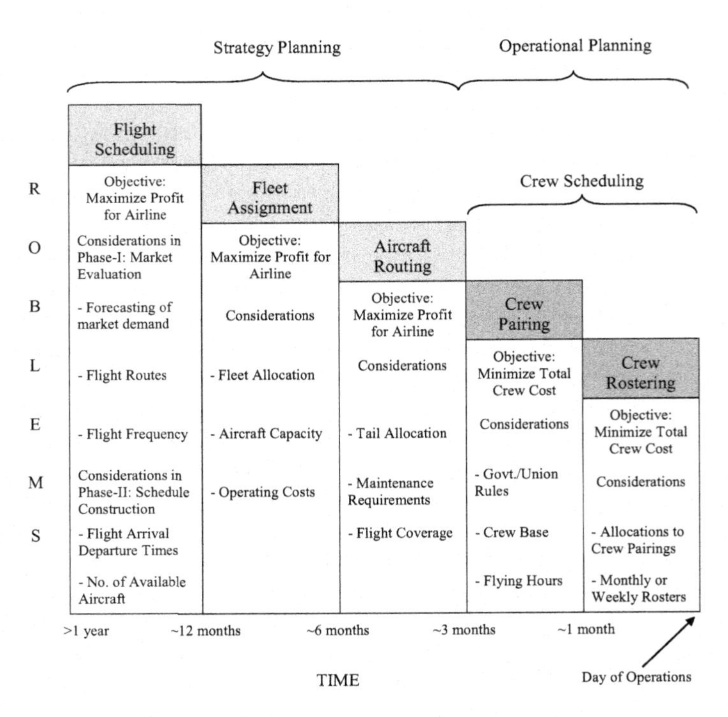

FIGURE 6.1 Sequential airline scheduling procedure.

6.2.2 Flight Schedule

An airline's decision to offer certain flights mainly depends on many factors such as seasonal demand forecasts, available fleet, market evaluation, pricing policies, and so on. Flight scheduling can be classified in different phases. The first phase corresponds to market evaluation where lists of routes, flight frequencies etc. are considered. This activity is conducted as early as two years before the day of operation. At this stage, the marketing department plays an important role in assessing the overall market scenario by taking marketing initiatives and competition information into account. In the second phase a schedule is obtained for the flights. This phase is around 12 months before the day of the flight. In this phase flight departure and arrival times, while satisfying resource constraints such as the number of available aircraft within a fleet and their overall maintenance requirements, are established.

Flight schedule or schedule design construction is considered as the first stage in airline planning and operations. A schedule is a timetable consisting of origin-destination pairs and departure/arrival times for flights that the airline intends to operate. A *flight leg* or *segment* is a nonstop flight between two ports. *Duration of flight* includes the block time, i.e., the time from aircraft engine start to engine shutdown and the time the flight is operational. The *operational time* of each leg is the difference between its actual arrival time and actual departure time. An important aspect in the planning phase is the consideration of *minimum turnaround time* i.e., the minimum time an aircraft must spend at the airport between two consecutive flights. It is also known as the *turnover duration* i.e., the time required after completion of a flight to prepare an aircraft for the next flight. Similarly, for crews, a connection between two flights is called a *sit connection.* The minimum sit connection time (30 min) requirement can be violated only if the crew follows the plane turn, i.e., it does not change planes. For details on technical aspects surrounding the aviation industry operations, refer to Appendix D.

6.2.3 Fleet Assignment

After the schedule design is completed, the next step is to assign a fleet type to each flight in the schedule. This process is called fleet assignment. Airlines generally operate with different fleet types, each having a different seating capacity, maintenance requirements, fuel consumption, and so on. The aim of fleet assignment is to maximize the profit by assigning the appropriate fleet type to the flights in the schedule (Barnhart et al., 2002). At this stage only a fleet type is allocated to a specific flight, not a particular aircraft. Fleet assignment is based on demand, operational costs, available seats in the aircraft and potential revenues. An airline's fleeting decision impacts its revenues and is an essential component of its overall scheduling process. Aircraft seats are an airline's perishable product i.e., unsold seats at the departure of the flight result in revenue loss to the airline. On the contrary, flights with fewer seats than demand may result in *spillover* costs. A spillover cost is the cost to the airline when the passenger chooses to avail their competitor's services which otherwise could have been provided to the customers with proper fleet allocation; eventually resulting in revenue loss to the airline. Therefore, the ideal strategy for the airline is to assign the appropriate aircraft fleet in the schedule to minimize unsold seats or spillover costs.

6.2.4 Aircraft Routing

Aircraft routing follows fleet assignment. It is a process of assigning each available aircraft within a fleet to specific flights. Aircraft routing is also referred to as aircraft assignment or tail assignment. The main objective of aircraft routing is to either maximize revenue or minimize operating costs for the airline. Important issues are considered while routing an aircraft, such as each flight must be covered by exactly one aircraft (*flight coverage*) and maintenance check conditions of the aircraft (*maintenance requirements*) are ensured. Aircraft maintenance is an important issue in airline operations. Airlines follow it strictly to ensure the safety of passengers, to abide by the laws of the aviation industry and to keep their asset (aircraft) in a good operating condition. For the airline, the aim is to ensure that the aircraft is available in operating condition as and when required. When a maintenance check is due for an aircraft, the aircraft must be present at the appropriate maintenance location because not all locations provide the facilities required for all the 'checks' and 'fleet' types. The Federal Aviation Administration (FAA) regulations state that the airlines are required to perform different types of aircraft maintenance checks, commonly referred to as A, B, C, and D checks. These checks vary in scope, duration, and frequency. Type A check (most frequent check) is carried out after every 65 flight hours, Type B check is performed between 300 and 600 of flight hours whereas Type C and D checks are carried out between one to four years of aircraft operations, respectively (Sriram & Haghani, 2003). Aircraft maintenance is an intricate process that requires state-of-the-art technology, advanced expertise, and considerable amount of capital to accomplish various checks. It is important to know that aircraft maintenance costs are approximately 12–15% of the total annual airline costs (Čokorilo, 2011). Therefore, aircraft should be assigned flights in the schedule such that they are available for the required maintenance check at the right time and at the right location.

6.2.5 Crew Scheduling

After aircraft routing is accomplished, the next stage is to do crew scheduling. Crew scheduling is a process of identifying sequences of flights and assigning both cockpit (technical) and cabin crew to these sequences by minimizing the cost of these assignments. Crew scheduling is completed 1–3 months prior to the actual day of operations, and it is one of the most computationally intensive combinatorial optimization problems (Ryan, 1992). Crew scheduling must satisfy constraints such as union, government and contractual regulations etc. Cockpit and cabin crews are governed by different sets of rules and various legal obligations such as restrictions on flying hours, duty periods and rest time between two duties, therefore their scheduling pattern is not identical. As explained, crew scheduling can further be segmented into two phases: crew pairing and crew assignment.

The first phase in crew scheduling is to generate crew pairings, also known as Tours of duty (ToD's). A crew pairing is a generic sequence of flights which can be operated by a crew from its crew base satisfying industry regulations and contractual restrictions (also called *legalities*). A crew base is a city where crew pairings begin or end. Crew bases are designated stations where crews are stationed which is often a *hub.* In this phase, cockpit and cabin crews are assigned to all flights for a specific period. For both these groups, individual flights are grouped to form pairings. A *duty* is a working period of a crew consisting of sequence of flights, where the arrival port of a flight is the departure port of the next flight. A duty is subject to regulatory (such as FAA) and company rules. When a crew is on duty, it flies a set of consecutive flight legs that follow certain regulations and contractual restrictions. A typical pairing consists of several *duties* which are preceded and followed by a rest period. However, at times, a sequence of flight legs may follow an overnight rest (*layover*) and may last up to a few days. In international sectors, a pairing may last as many as seven days. Like sit connection times, there is a lower and an upper bound on the rest, denoted by *minimum rest (30 min)* and *maximum rest (varies)*, respectively. An illustration of a ToD is shown in Figure 6.2 in which a ToD consists of five flying tasks to be operated by a crew. The flying tasks are followed by a rest period. IATA airport codes are used to depict the destinations in the ToD.

After boarding the aircraft, in the check-in duration phase, both the technical and cabin crews carry out the necessary checks and preparations required to ensure that the aircraft is airworthy. Similarly, after the landing and reaching the apron, crew carry out the necessary post-flight inspections, before leaving the aircraft (Wilke et al., 2014).

It is important to know that aircraft fuel costs and crew costs are the two most significant expenses for an airline and are accounted for during the planning stages of airline operations (Ball et al., 2006). The fuel and crew costs amount to approximately 20 percent of the total operational costs for an airline (Gopalakrishnan & Johnson, 2005). Total crew cost is a combination of linear costs (salaries, hotels) and non-linear costs (actual flying time, total work time). However, unlike fuel costs most crew costs can be controlled by proper scheduling. Therefore, considerable emphasis is given to minimize the crew costs. Moreover, crew costs occur regardless the crew is flying or paxing. *Paxing* or *deadheading* is a situation when crew members travel as passengers to be available on a flight that does not depart from their current location

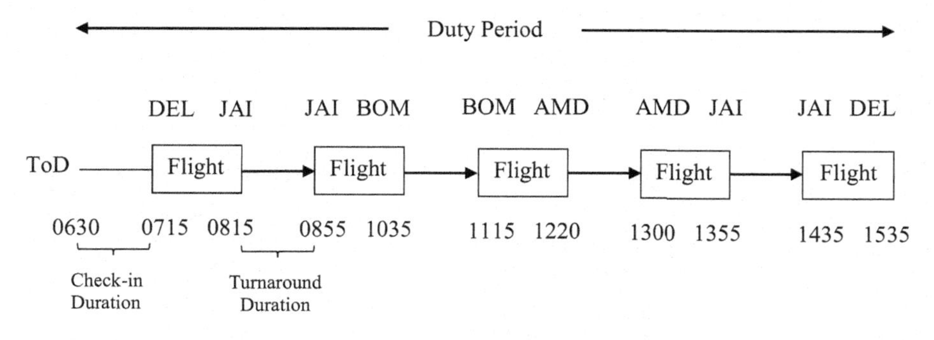

FIGURE 6.2 A Tour of duty for a crew.

Day	1	2	3	4	5	6	7	8	9	10	11	12	13	14
Crew	Fly	Fly	Fly	Fly	CALL	Fly	O	Fly	Fly	SIM	Fly	Fly	CALL	Fly

FIGURE 6.3 A two-week roster for a crew.

or to return to their respective crew base. The objective of crew pairing is to find a set of legal pairings that cover all flights and minimize the total crew cost. Crew pairing problems can be placed in three categories: the daily problem, the weekly problem, and the dated problem. In the *daily problem* the same flights are assumed to be flown every day i.e., the schedule is repeated on the daily basis. The *weekly problem* assumes that the schedule would repeat itself every week. However, the time horizon for the *dated problem* is usually longer and it must be solved when shifting from one schedule to another for solutions that spans in both schedules.

The second phase in crew scheduling is called crew rostering or crew assignment. In this phase crew pairings are gathered into work schedules and assigned to specific crew members. Depending on the airline's approach, either *crew rostering* or *preferential bidding* is used for crew assignment. Generally, European airlines generate crew rosters whereas airlines in North America follow the preferential bidding approach. Crew rostering is the process of assigning individual crew members to crew pairings based upon the individual's preferences. The airline then attempts to grant these rosters, also known as *lines of work*. Crew assignments can be made on a monthly, fortnightly, or weekly basis. In preferential bidding, crew members bid on their relative preferences and the airline assigns specific schedules to crew members based upon their bids and seniority. Figure 6.3 shows a two-week roster for a crew where 'Fly' represents a flying duty in the ToD, 'CALL' means on-call duty, 'O' represents a day off and 'SIM' means a simulator session for the crew.

6.3 INDIA-BASED DOMESTIC NETWORK

In India, domestic air network is developing fast and connects major cities through multiple daily flights and are also well connected with relatively small centers with varying flight frequency. The network operates with major hub cities such as Delhi (DEL), Mumbai (BOM), Chennai (MAA), and Kolkata (CCU). These hubs are connected with spoke cities such as Jaipur (JAI), Ahmedabad (AMD), Lucknow (LKO), or Srinagar (SXR). However, due to increasing traffic, there is also direct connectivity of non-hub cities in the domestic and international sectors.

6.3.1 Flight Network

Based on the domestic air routes in India, a network of flights connecting two hub (DEL and BOM) and two spoke (JAI and AMD) cities is generated to illustrate the concepts of aircraft routing and crew scheduling as coved in the previous sections is presented. Therefore, first, indicative flight durations between four airports in India

TABLE 6.1
Flight durations between the hub and spoke cities included in the network

Delhi (DEL)			
Jaipur (JAI)	1 h		
Ahmedabad (AMD)	1 h 20 min	45 min	
Mumbai (BOM)	1 h 50 min	1 h 40 min	1 h
	Delhi	Jaipur	Ahmedabad

are presented in Table 6.1. However, the actual flight duration may differ as they depend on factors such as the type of aircraft used, the flying altitude and wind speed encountered during the flight.

Based on the flight durations between the four cities, a flight network consisting of 23 flights is presented in Figure 6.4. In this network, a node represents a port and an arc represents a flight. Every arc has flight properties such as flight number, departure city and departure time, arrival city and arrival time, aircraft-, captain- and first officer-identifiers. Based on this network, airline scheduling aspects are presented in the following sections.

6.3.2 Aircraft Routings

Based on the flight network presented in Figure 6.4, four aircraft routings are generated to cover all flights in the network. The aircraft routings are color coded in red, yellow, green, and blue to distinguish the flights each aircraft is assigned. For simplicity, the flights are labeled numerically in each color scheme giving each flight a unique label. Therefore, aircraft identifiers considered are generic and do not resemble the operations of any commercial airline.

From the schedule in Figure 6.4, aircraft routings and the turnaround time for each aircraft operating in the network can be established. Aircraft routings are generated with the assumption that all mandatory aircraft checks for a single aircraft fleet conform to maintenance regulations. Furthermore, in each routing, a minimum 30-min ground turnaround time between two consecutive flights is ensured to reflect real-world operations.

Based on Figure 6.4, aircraft routings are presented in Table 6.2 through to Table 6.5.

Based on aircraft routings presented in Table 6.2, the ground time or turnaround duration for aircraft A1 between any two consecutive flights in the network can be identified. For instance, the first flight of the day for aircraft A1 i.e., flight 1 departs from DEL at 0715 and arrives in JAI at 0815, whereas the next scheduled operation for aircraft A1 i.e., flight 2 departs from JAI at 0855 leaving a ground time of 40 min for aircraft A1 at JAI. Similarly, for all subsequent flights on the day of operations, aircraft A1 has an identical ground time of 40 min in the network.

Based on aircraft routings presented in Table 6.3, the ground time or turnaround duration for aircraft A2 between any two consecutive flights in the network can also

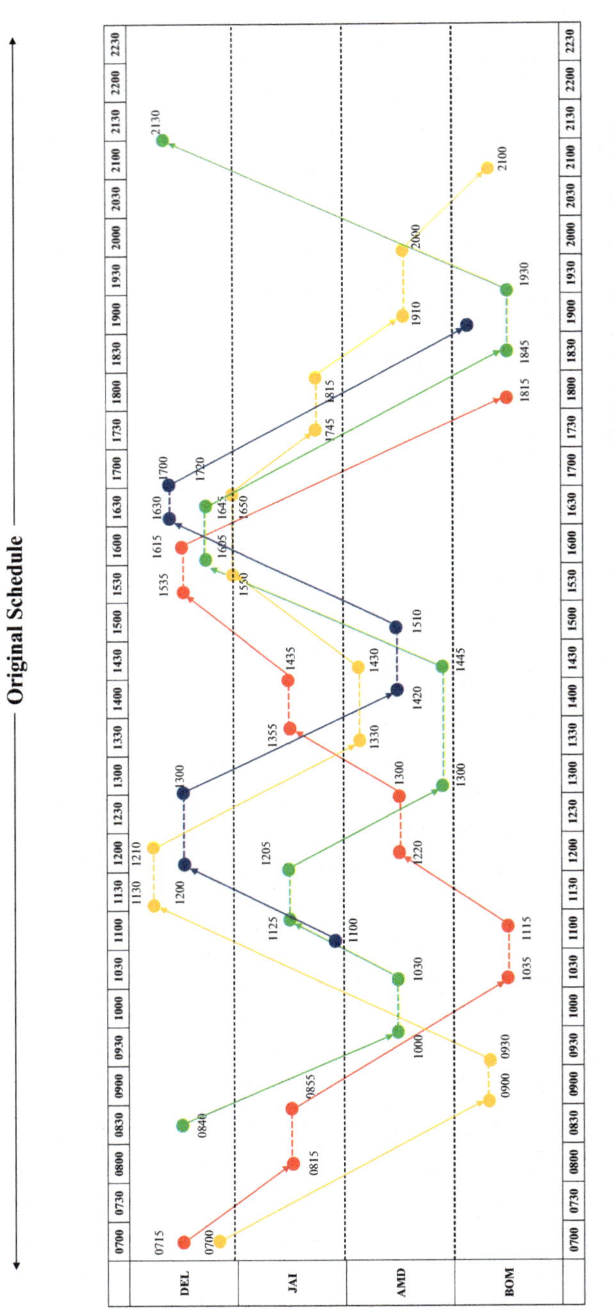

FIGURE 6.4 Planned domestic network of an airline in India covering four cities through 23 flights using four aircraft and 12 crews.

TABLE 6.2
Routings for aircraft A1 (red)

Flight 1		Flight 2		Flight 3		Flight 4		Flight 5		Flight 6	
DEL	JAI	JAI	BOM	BOM	AMD	AMD	JAI	JAI	DEL	DEL	BOM
0715	0815	0855	1035	1115	1220	1300	1355	1435	1535	1615	1815

TABLE 6.3
Routings for aircraft A2 (yellow)

Flight 1		Flight 2		Flight 3		Flight 4		Flight 5		Flight 6		Flight 7	
DEL	BOM	BOM	DEL	DEL	AMD	AMD	DEL	DEL	JAI	JAI	AMD	AMD	BOM
0700	0900	0930	1130	1210	1330	1430	1550	1650	1745	1815	1910	2000	2100

TABLE 6.4
Routings for aircraft A3 (green)

Flight 1		Flight 2		Flight 3		Flight 4		Flight 5		Flight 6	
DEL	AMD	AMD	JAI	JAI	AMD	AMD	DEL	DEL	BOM	BOM	DEL
0840	1000	1030	1125	1205	1300	1445	1605	1645	1845	1930	2130

be identified. The first flight of the day for aircraft A2 i.e., flight 1 departs from DEL at 0700 and arrives in BOM at 0900. After this, the next scheduled task for aircraft A2 i.e., flight 2 departs from BOM at 0930 leaving a ground time of 30 min for aircraft A2 at BOM. The ground time for aircraft A2 before the subsequent flights in the network i.e., flights 3, 4, 5, 6, and 7 is 40 min (at DEL), 60 min (at AMD), 60 min (at DEL), 30 min (at JAI), and 50 (AMD), respectively.

Based on aircraft routings presented in Table 6.4, the ground time for aircraft A3 between any two consecutive flights in the network can be found. The first flight of the day for aircraft A3 i.e., flight 1 departs from DEL at 0840 and arrives in AMD at 1000. After this, the next scheduled operation for aircraft A2 i.e., flight 2 departs from AMD at 1030 leaving a ground time of 30 min for aircraft A2 at AMD. The ground time for aircraft A3 before the remaining scheduled flights in the network i.e., flights 3, 4, 5, and 6 is 40 min (at JAI), 105 min (at AMD), 40 min (at DEL), and 45 min (at BOM), respectively.

Based on aircraft routings presented in Table 6.5, the ground time for aircraft A4 between any two consecutive flights in the network can be ascertained. The first flight of the day for aircraft A4 i.e., flight 1 departs from JAI at 1100 and arrives in DEL at 1200. The next scheduled task for aircraft A4 i.e., flight 2 departs from DEL at 1300

TABLE 6.5
Routings for aircraft A4 (blue)

Flight 1		Flight 2		Flight 3		Flight 4	
JAI	DEL	DEL	AMD	AMD	DEL	DEL	BOM
1100	1200	1300	1420	1510	1630	1700	1900

TABLE 6.6
Captain C1 pairings (red)

Flight 1		Flight 2		Flight 3		Flight 4		Flight 5	
DEL	JAI	JAI	BOM	BOM	AMD	AMD	JAI	JAI	DEL
0715	0815	0855	1035	1115	1220	1300	1355	1435	1535

leaving a ground time of 60 min for aircraft A4 at DEL. The ground time for aircraft A4 before the remaining scheduled flights i.e., flights 3 and 4 is 50 min (at AMD) and 30 min (at DEL), respectively.

6.3.3 Crew Pairings

After creating the aircraft routings, captain and first officer pairings are generated. Generally, the captain and the first officer stay together i.e., they perform same flight tasks. However, in certain instances, captain and first officer may split but still end the ToDs at their respective crew bases. Based on Figure 6.4, crew pairings for the captains and the first officers are presented in Table 6.6 through to Table 6.17.

Based on crew pairings presented in Table 6.6, the sit time for captain C1 between the two consecutive flights in the scheduled network can be identified. The first flight of the day for captain C1 i.e., flight 1 departs from DEL at 0715 and arrives in JAI at 0815. The next scheduled task for captain C1 i.e., flight 2 departs from JAI at 0855 leaving a sit time of 40 min for captain C1 at JAI. Therefore, the sit time for captain C1 before the remaining scheduled flights i.e., flights 3, 4, and 5 is 40 min (at BOM), 40 min (at AMD), and 40 min (at JAI), respectively. As evident from Tables 6.6 and 6.2, Captain C1 is scheduled to fly aircraft A1 and stays with it till the completion of flight 5 in the scheduled network.

Based on crew pairings presented in Table 6.7, the sit time for captain C2 between the two consecutive flights in the scheduled network can be identified. The first flight of the day for captain C2 i.e., flight 1 departs from DEL at 0700 and arrives in BOM at 0900. The next scheduled task for captain C2 i.e., flight 2 departs from BOM at 0930 leaving a sit time of 30 min for captain C2 at BOM. Therefore, the sit time for captain C2 before the remaining scheduled flights i.e., flights 3 and 4 is 40 min (at DEL) and 60 min (at AMD), respectively. As evident from Tables 6.7 and 6.3, Captain

TABLE 6.7
Captain C2 pairings (yellow)

Flight 1		Flight 2		Flight 3		Flight 4	
DEL	BOM	BOM	DEL	DEL	AMD	AMD	DEL
0700	0900	0930	1130	1210	1330	1430	1550

TABLE 6.8
Captain C3 pairings (green)

Flight 1		Flight 2		Flight 3		Flight 4	
DEL	AMD	AMD	JAI	JAI	AMD	AMD	DEL
0840	1000	1030	1125	1205	1300	1445	1605

TABLE 6.9
Captain C4 pairings (blue)

Flight 1		Flight 2		Flight 3		Flight 4	
JAI	DEL	DEL	AMD	AMD	DEL	DEL	BOM
1100	1200	1300	1420	1510	1630	1700	1900

C2 is scheduled to fly aircraft A2 and stays with it till the completion of flight 4 in the scheduled network.

Based on crew pairings presented in Table 6.8, the sit time for captain C3 between the two consecutive flights in the scheduled network can be identified. The first flight of the day for captain C3 i.e., flight 1 departs from DEL at 0840 and arrives in AMD at 1000. The next scheduled task for captain C3 i.e., flight 2 departs from AMD at 1030 leaving a sit time of 30 min for captain C3 at AMD. Therefore, the sit time for captain C3 before the remaining scheduled flights i.e., flights 3 and 4 is 40 min (at JAI) and 105 min (at AMD), respectively. As evident from Tables 6.8 and 6.4, Captain C3 is scheduled to fly aircraft A3 and stays with it till the completion of flight 4 in the scheduled network.

Based on crew pairings presented in Table 6.9, the sit time for captain C4 between the two consecutive flights in the scheduled network can be identified. The first flight of the day for captain C4 i.e., flight 1 departs from JAI at 1100 and arrives in DEL at 1200. The next scheduled task for captain C4 i.e., flight 2 departs from DEL at 1300 leaving a sit time of 60 min for captain C4 at DEL. Therefore, the sit time for captain C4 before the remaining scheduled flights i.e., flights 3 and 4 is 50 min (at AMD) and 30 min (at DEL), respectively. As evident from Table 6.9 and 6.5, Captain C4

TABLE 6.10
Captain C5 pairings (yellow)

Flight 5		Flight 6		Flight 7	
DEL	JAI	JAI	AMD	AMD	BOM
1650	1745	1815	1910	2000	2100

TABLE 6.11
Captain C6 pairings (green)

Flight 5		Flight 6	
DEL	BOM	BOM	DEL
1645	1845	1930	2130

TABLE 6.12
First Officer F1 pairings (red F1, F2, and F3; yellow F4)

Flight 1		Flight 2		Flight 3		Flight 4	
DEL	JAI	JAI	BOM	BOM	AMD	AMD	DEL
0715	0815	0855	1035	1115	1220	1430	1550

is scheduled to fly aircraft A4 and stays with it till the completion of flight 4 in the scheduled network.

Based on crew pairings presented in Table 6.10, the sit time for captain C5 between the two consecutive flights in the scheduled network can be identified. The first flight of the day for captain C5 i.e., flight 5 departs from DEL at 1650 and arrives in JAI at 1745. The next scheduled task for captain C5 i.e., flight 6 departs from JAI at 1815 leaving a sit time of 30 min for captain C5 at JAI. Therefore, the sit time for captain C5 before the remaining scheduled flight i.e., flight 7 50 min (at AMD). As evident from Tables 6.10 and 6.3, Captain C5 is scheduled to fly aircraft A2 and stays with it till the completion of flight 7 in the scheduled network.

Based on crew pairings presented in Table 6.11, the sit time for captain C6 between the two consecutive flights in the scheduled network can be identified. The first flight of the day for captain C6 i.e., flight 5 departs from DEL at 1645 and arrives in BOM at 1445. The next scheduled task for captain C6 i.e., flight 6 departs from BOM at 1930 leaving a sit time of 45 min for captain C6 at BOM. As evident from Tables 6.11 and 6.4, Captain C6 is scheduled to fly aircraft A3 and stays with it till the completion of flight 6 in the scheduled network.

Based on crew pairings presented in Table 6.12, the sit time for first officer F1 between the two consecutive flights in the scheduled network can be identified.

The first flight of the day for first officer F1 i.e., flight 1 departs from DEL at 0715 and arrives in JAI at 0815. The next scheduled task for first officer F1 i.e., flight 2 departs from JAI at 0855 leaving a sit time of 40 min for first officer F1 at JAI. Therefore, the sit time for first officer F1 before the remaining scheduled flights i.e., flights 3 and 4 is 40 min (at BOM) and 130 min (at AMD), respectively. As evident from Tables 6.12, 6.2, and 6.3, first officer F1 is scheduled to fly aircraft A1 (flights 1, 2, and 3) and aircraft A2 (flight 4) in the scheduled network. It is important to know that first officer F1 arrives in AMD with aircraft A1 and captain C1 at 1220 and departs AMD at 1430 with aircraft A2 and Captain C2 (Table 6.7). Mostly, a captain and first officer fly together, however in some instances due to a certain set of constraints the captain and first officer split and end their pairings at their respective crew bases.

Based on crew pairings presented in Table 6.13, the sit time for first officer F2 between the two consecutive flights in the scheduled network can be identified. The first flight of the day for first officer F2 i.e., flight 1 departs from DEL at 0700 and arrives in BOM at 0900. The next scheduled task for first officer F2 i.e., flight 2 departs from BOM at 0930 leaving a sit time of 30 min for first officer F2 at BOM. Therefore, the sit time for first officer F2 before the remaining scheduled flights i.e., flights 3 and 4 is 40 min (at DEL) and 75 min (at AMD), respectively. As evident from Tables 6.13, 6.3, and 6.4, first officer F2 is scheduled to fly aircraft A2 (flights 1, 2, and 3) and aircraft A3 (flight 4) in the scheduled network. First officer F2 arrives in AMD with aircraft A2 and captain C2 at 1330 and departs AMD at 1445 with aircraft A3 and Captain C3 (Table 6.8).

Based on crew pairings presented in Table 6.14, the sit time for first officer F3 between the two consecutive flights in the scheduled network can be identified. The first flight of the day for first officer F3 i.e., flight 1 departs from DEL at 0840 and arrives in AMD at 1000. The next scheduled task for first officer F3 i.e., flight 2

TABLE 6.13
First Officer F2 pairings (yellow F1, F2, and F3; green F4)

Flight 1		Flight 2		Flight 3		Flight 4	
DEL	BOM	BOM	DEL	DEL	AMD	AMD	DEL
0700	0900	0930	1130	1210	1330	1445	1605

TABLE 6.14
First Officer F3 pairings (green F1, F2, and F3; blue F3 & F4)

Flight 1		Flight 2		Flight 3		Flight 3		Flight 4	
DEL	AMD	AMD	JAI	JAI	AMD	AMD	DEL	DEL	BOM
0840	1000	1030	1125	1205	1300	1510	1630	1700	1900

departs from AMD at 1030 leaving a sit time of 30 min for first officer F3 at JAI. Therefore, the sit time for first officer F3 before the remaining scheduled flights i.e., flights 3, 4, and 5 is 40 min (at JAI), 130 min (at AMD), and 30 min (at DEL), respectively. As evident from Tables 6.14, 6.4, and 6.5, first officer F3 is scheduled to fly aircraft A3 (flights 1, 2, and 3) and aircraft A4 (flights 3 and 4) in the scheduled network. First officer F3 arrives in AMD with aircraft A3 and captain C3 at 1300 and departs AMD at 1510 with aircraft A4 and Captain C4 (Table 6.9).

Based on crew pairings presented in Table 6.15, the sit time for first officer F4 between the two consecutive flights in the scheduled network can be identified. The first flight of the day for first officer F4 i.e., flight 1 departs from JAI at 1100 and arrives in DEL at 1200. The next scheduled task for first officer F4 i.e., flight 2 departs from DEL at 1300 leaving a sit time of 60 min for first officer F4 at DEL. As evident from Tables 6.15, 6.5, and 6.9, first officer F4 is scheduled to fly aircraft A4 (flights 1 and 2) in the scheduled network. First officer F4 starts the duty in JAI with aircraft A4 and captain C4 (Table 6.9).

Based on crew pairings presented in Table 6.16, the sit time for first officer F5 between the two consecutive flights in the scheduled network can be identified. The first flight of the day for first officer F5 i.e., flight 4 departs from AMD at 1300 and arrives in JAI at 1355. The next scheduled task for first officer F5 i.e., flight 5 departs from JAI at 1435 leaving a sit time of 40 min for first officer F5 at JAI. Therefore, the sit time for first officer F5 before the remaining scheduled flights i.e., flight 6 (of A1) and flight 6 (of A3) is 40 min (at DEL) and 75 min (at BOM), respectively. As evident from Tables 6.16, 6.2, and 6.4, first officer F5 is scheduled to fly aircraft A1 (flights 4, 5, and 6) and aircraft A3 (flight 6) in the scheduled network. First officer F5 arrives in BOM from DEL with aircraft A1 and a reserve captain (rare instance) at 1815 and departs BOM at 1930 with aircraft A3 and Captain C6 (Table 6.11).

TABLE 6.15
First Officer F4 pairings (blue F1 and F2)

Flight 1		Flight 2	
JAI	**DEL**	**DEL**	**AMD**
1100	1200	1300	1420

TABLE 6.16
First Officer F5 pairings (red F4, F5, and F6; green F6)

Flight 4		Flight 5		Flight 6		Flight 6	
AMD	**JAI**	**JAI**	**DEL**	**DEL**	**BOM**	**BOM**	**DEL**
1300	1355	1435	1535	1615	1815	1930	2130

TABLE 6.17
First Officer F6 pairings (yellow F5, F6, and F7)

Flight 5		Flight 6		Flight 7	
DEL	JAI	JAI	AMD	AMD	BOM
1650	1745	1815	1910	2000	2100

Based on crew pairings presented in Table 6.17, the sit time for first officer F6 between the two consecutive flights in the scheduled network can be identified. The first flight of the day for first officer F6 i.e., flight 5 departs from DEL at 1650 and arrives in JAI at 1745. The next scheduled task for first officer F6 i.e., flight 6 departs from JAI at 1815 leaving a sit time of 30 min for first officer F6 at JAI. Therefore, the sit time for first officer F6 before the remaining scheduled flight i.e., flight 7 is 50 min (at AMD). As evident from Tables 6.17, 6.3, and 6.10, first officer F6 is scheduled to fly aircraft A2 (flights 5, 6, and 7) with captain C5 in the scheduled network (Table 6.10).

6.4 OTHER SCHEDULING TECHNIQUES

Apart from the traditional sequential approach of airline scheduling there are substantial attempts made to integrate various components of the airline scheduling problem. The sequential problems of schedule design, fleet assignment, aircraft routing, and crew scheduling have interdependencies on one another. Therefore, the optimal solution for these problems considered separately may not yield a solution that is optimal for the integrated problem. On the contrary, an integrated approach to address these problems may result in increased revenue, improved flight connection opportunities, cost savings, and so on.

Schedule disruptions (covered in Chapters 7 and 8) are unavoidable and tend to happen frequently, especially the minor ones. To overcome these unwanted and often problematic glitches in the schedule, the concept of robustness is incorporated while designing airline schedules. Robust airline scheduling is an approach to minimize schedule disruptions by making the airline schedule less vulnerable to disruptions. In other words, robust schedules can accommodate disruptions within the schedule itself. Robustness can be broadly categorized into two areas; *absorption robustness* and *recovery robustness* (Clausen et al., 2010). Absorption robustness maintains the feasibility of the schedule during smaller disruptions and absorbs the ripple effects of the disruption by adding time buffers into the schedule. Recovery robustness looks at designing aircraft rotations and crew schedules in a manner that recovery plans such as aircraft and crew swaps, delaying and cancelling flights are feasible during disruptions. However, incorporating robustness into schedules is a difficult task because of the unknown severity level of the disruptions and the complexity associated in estimating the costs related with robust scheduling.

6.5 CONCLUSION

In airline operations, the primary focus of competition between different operators remains cost and responsiveness. To gain a competitive advantage they employ operational strategies for airline scheduling which are both expensive (time, cost) and expansive (resource utilization). Airline schedule construction via a sequential scheduling procedure is critical for airline operations and it entails solving fleet assignment, aircraft routing, crew pairing, and rostering problems. Besides this, other scheduling aspects such as robust and integrated scheduling are also considered to gain efficiencies in the operations. To elaborate upon the dimensions of aircraft routing and crew paring, a domestic airline network is presented highlighting the airline schedule construction aspects.

CHAPTER QUESTIONS

Q1. Based on the regulations of generating the crew pairings, identify and discuss the limitations, if any, in the captain crew pairings presented in Table 6.6 through to Table 6.11.

Q2. Generate an alternate set of crew pairings for each captain to cover the flights in the network.

Q3. Is any flight uncovered by a captain? If so, identify the flight and assign a captain ensuring crew allocation legalities.

Q4. Based on the regulations of generating the crew pairings, identify and discuss the limitation, if any, in the first officer crew pairings presented in Table 6.12 through to Table 6.17.

Q5. Generate an alternate set of crew pairings for each first officer to cover the flights in the network.

Q6. Is any flight uncovered by a first officer? If so, identify the flight and assign a first officer ensuring crew allocation legalities.

Q7. Discuss the difference between sequential airline scheduling, integrated scheduling and robust scheduling approaches. What are the benefits and limitations of each scheduling approach?

REFERENCES

Ball, M. O., Barnhart, C., Nemhauser, G., & Odoni, A. (2006). Air transportation: Irregular operations and control. In C. Barnhart & G. Laporte, (Eds), *Handbook of Operations Research and Management Science: Transportation*, Elsevier.Barnhart, C., Kniker, T. S., & Lohatepanont, M. (2002). Itinerary-based airline fleet assignment. *Transportation Science*, *36*(2), 199–217.

Barnhart, C., & Talluri, K. T. (1997). Airlines operations research. In C. ReVelle & A. McGraity (Eds), *Design and Operation of Civil and Environmental Engineering System*. pp. 435–469, Wiley.

Clausen, J., Larsen, A., Larsen, J., & Rezanova, N. J. (2010). Disruption management in the airline industry—concepts, models and methods, *Computers and Operations Research*, *37*(5), 809–821.

Čokorilo, O. (2011). Aircraft performance: The effects of the multi attribute decision making of non-time dependent maintainability parameters. *International Journal for Traffic and Transport Engineering*, *1*(1), 41–47.

Gopalakrishnan, B., & Johnson, E. L. (2005). Airline crew scheduling: State-of-the-art. *Annals of Operations Research*, *140*, 305–337.

Padfield, G.D. & Lawrence, B. (2003). The birth of flight control: An engineering analysis of the Wright Brothers' 1902 glider, *The Aeronautical Journal*, *107* (1078), 697–718.

Ryan, D. M. (1992). The solution of massive, generalized set partitioning problems in aircrew rostering. *Journal of Operations Research Society*, *43*(5), 459–467.

Sriram, C., & Haghani, A. (2003). An optimization model for aircraft maintenance scheduling and re-assignment. *Transportation Research Part A*, *37*, 29–48.

Wilke, S., Majumdar, A., & Ochieng, W. Y. (2014). Airport surface operations: A holistic framework for operations modeling and risk management. *Safety Science*, *63*, 18–33.

7 Airline Disruption Management – I

CHAPTER OBJECTIVES

At the end of this chapter, you will be able to

- Understand the disruption management in airline operations.
- Understand the concept of airline schedule recovery.
- Understand various aircraft recovery (ARO) approaches used during disruptions.
- Understand different crew recovery aspects considered in the airline operations.
- Know about passenger and integrated recovery aspects.

7.1 INTRODUCTION

In this chapter, airline disruption management aspects are presented. Schedule recovery in the airline industry has been an area of interest for researchers and practitioners for the last couple of decades. A schedule is recovered during disruptions by applying recovery approaches such as delaying and/or canceling flights at different stages of the recovery process. A flight normally operates with resources such as an aircraft, technical crew (pilot and first officer), and cabin crew (flight attendants). During schedule recovery, aircraft and crews – the two most significant and expensive airline resources are the focus of the recovery operations. Therefore, during disruption management, airlines try to come out of perturbed schedule as early as possible to minimize costs and inconvenience to passengers by making the optimal use of their resources.

7.2 DISRUPTION MANAGEMENT

The aviation industry has seen substantial changes in the last couple of decades. Over the years air traffic has augmented considerably whereas many airlines fail to proportionately increase their resources such as aircraft and crew to cater to the ever-increasing customer demand. This leads to the highly optimized utilization of the resources in the schedule. In other words, airlines develop optimal schedules leaving very little slack to deal with unplanned operations. A s*lack* is the time period when a resource is not assigned a task or duty after exhausting mandatory ground time. From the planning perspective, the slack in the schedule is not considered useful; indicating

DOI: 10.1201/9780203731338-7

underutilization of a resource, however from the operational view it is helpful as it gives time to absorb or wash-out disruptions rather than allowing it to propagate in the network (Appendix B). The tight coupling of resources often results in operational problems for the airlines – especially during disruptions.

Throughout the planning process it is assumed that an airline schedule would be implemented as planned without taking unforeseen events such as disruptions into consideration. However, disruptions do occur during operations and cause problems for the airline and passengers. Disruptions are of different kinds and their level of impact on the schedule varies. Disruptions of short duration (5 to 10 min) have less or no impact on the schedule than those that last an hour or more. In short haul flights, a delay of an hour or so has a significant effect on airline schedules and operations. The reasons for disruption in the schedule are inclement weather, unscheduled maintenance, unavailability of crew and/or aircraft, passenger delay, security issues, airspace congestion (AhmadBeygi et al., 2010). Airline schedule recovery i.e., getting back to normal operations has many challenges. In case of disruptions in the schedule, the aim is to bring the disrupted schedule back to normal operations as soon as possible by re-allocating the affected flights in the schedule in a feasible and cost-effective manner. When disruptions occur, airlines adjust their flight operations by delaying flight departures, canceling flights, rerouting aircraft, reassigning crew, and re-accommodating passengers.

7.3 TYPES OF DISRUPTION

Schedule disruptions can be broadly classified as ground (Type-I) and mid-air (Type-II) disruptions.

Type-I: Ground Disruption

This type of disruption is more common in day-to-day operations. Ground disruptions happen when a flight is not able to take off as scheduled and is delayed. The major reasons for this disruption include, but are not limited to; technical problems, unavailability of aircraft and/or crew, passenger delay, security reasons, and so on. In this case, disruption is acknowledged at the departure port of the delayed flight.

Type-II: Mid-air Disruption

Mid-air disruptions occur when a flight is airborne. Frequent reasons for this type of disruption are, but are not limited to, convective weather and airspace congestion. In this type of disruption, time of disruption is critical as it would determine the actual port of arrival for the disrupted flight. The decision where to land the disrupted flight is taken between the Air Traffic Control (ATC) and the technical crew of the disrupted flight. However, in general the prerogative lies with the ATC to ensure the safe landing of the aircraft. Broadly, the identification for the arrival port for the disrupted flight can be classified based on the follow scenarios.

Case-I: Arrival at the Scheduled Port of Arrival

This situation arises if disruption is known when the disrupted flight is approaching close to its scheduled port of arrival. In this case, neighborhood is generated at the arrival port of the disrupted flight.

Case-II: Arrival at the Port of Departure

However, if the time of disruption is close to the departure time of the disrupted flight, then the flight may come back to the port of departure instead of proceeding further. In this case, port of disruption would be the departure port of the disrupted flight and disruption neighborhood will be generated at this port.

Case-III: Arrival at Some Other Port

Also, a disrupted flight may land on a port which is neither its departure nor arrival port. In this situation the new arrival port will be the port at which resources will be generated for neighborhood creation.

Regardless of the disruption type, a significant delay in flight departure time (*on-ground delays*) or flight arrival time (*mid-air delays*) may affect the aircraft, crews, and passengers. Early day disruptions are to be resolved as soon as possible since later in the day the delay may propagate through the network and affect other flights, eventually leading to the breakdown of the originally planned schedule. Especially, in short haul operations, disruptions that occur early in a day tend to have more adverse impact on the schedule than those that happen late in the evening because most of the short haul flights operate during daytime.

Once disruption is observed in the airline schedule, then it is required to get the disrupted schedule back to normal operations. A disruption in the schedule may cause significant ripple effects in the flight network by affecting other flights and their resources; eventually leading to the breakdown of the planned schedule. Moreover, a disruption in the schedule may affect the network to the extent that flights which are not directly affected by the disrupted flight may also get disrupted by one way or the other. A disrupted flight is one, which does not depart or does not arrive on time.

Consider an example (Figure 7.1) at an airport where flight A (which is scheduled to arrive at 0900 h) is delayed by an hour because of a technical snag at its port of origin and arrives at 1000 h. This delay affects flight B which is scheduled to depart at 0930 h by the aircraft of flight A, the captain of flight C (which has already arrived as scheduled at 0800 h) and the first officer who starts its duty at 0930 h. Hence, due to the delayed arrival of the aircraft of flight A, flight B is also delayed and can be re-scheduled to depart at a later time (say at 1100 h). The captain of flight A was scheduled to do flight E at 1030 h but flight E cannot depart on time due to the late arrival of the captain of flight A and as a result flight E has to be re-scheduled with a later departure time (say at 1200 h). Also, the first officer of flight A is scheduled to operate flight D at 1000 h with the aircraft of flight C and along with the captain who starts its day of operations at 1000 h. Due to the unavailability of the first officer of flight A, flight D is also delayed and can be re-scheduled later in the day (say at 1130 h). Thus, the original delay of flight A causes flights B, D, and E to be delayed. Later in the day the delay may propagate through the network and affect other flights, eventually leading to the breakdown of the originally planned schedule.

An illustration of the above example is shown in Figure 7.1 where solid blue, red, and green arrows represent scheduled arrival or departure of aircraft, captain, and first officer, respectively. The dotted arrows represent actual arrival or departure times of these resources.

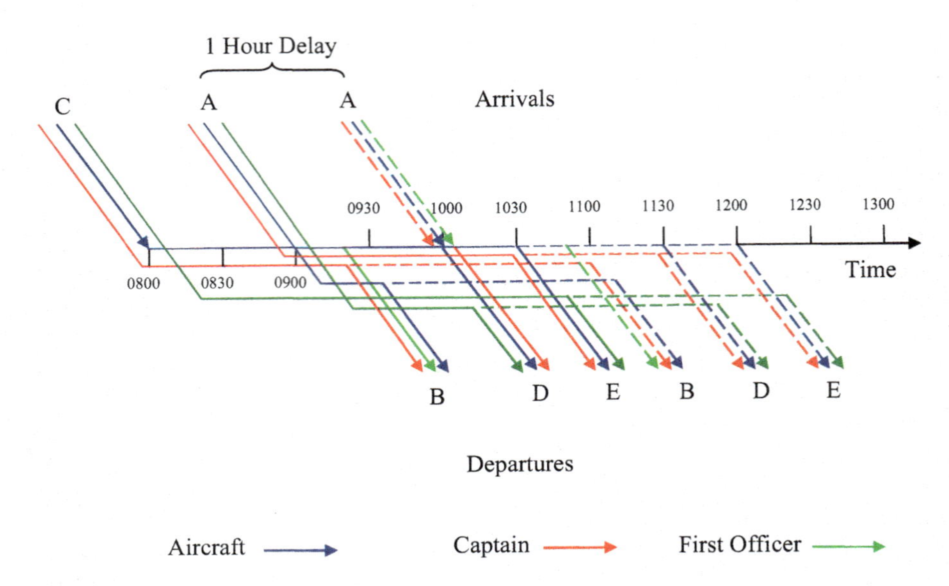

FIGURE 7.1 A disrupted airline schedule.

7.4 SCHEDULE RECOVERY FROM DISRUPTIONS

A delayed flight affects other flights (in case of substantial delays) in more than one aspect and may result in broken: aircraft routings, crew pairings, and passenger itineraries. As a result, an aircraft and crew may not operate a flight as scheduled and may require a set of new duties.

Schedule disruptions result in undesirable expenses to the airline, passenger inconvenience and dissatisfaction, disturbed aircraft maintenance checks, and so on. The Airline Operations Control Center (AOCC) plays an important role in the schedule recovery. The AOCC monitors flight irregularities and implements recovery plans. In case of severe disruptions, operations controllers may let an aircraft fly without passengers (*ferrying*), to a different airport (*diverting*), or toward another scheduled destination (*over-flying*). A recovery schedule should be constructed such that all aircraft and crew will be at appropriate ports by certain times to recover the schedule in a specified time period. However, a specific aircraft would not be considered for a given sequence of flights unless it meets its maintenance requirements; that is, it must be delivered to a maintenance location within the remaining legal flying time. The sequential decision process, first aircraft, then crew, and finally passenger recovery, is reflected in the research on airline recovery (Ball et al., 2007). Since airline schedule recovery is associated with many conflicting objectives hence quality of recovery solutions is not easy to determine.

During recovery, changes in aircraft routes may affect crew Tours of Duty (ToD)'s as planned ToD's may not be feasible anymore. Other reasons for infeasibility of ToD's could be missed connections, lack of available flying time or flight cancellations. The objective of the airline crew re-scheduling problem is to find a minimum cost reassignment of crews to a disrupted flight schedule, taking into consideration flown hours of the crews, partially flown pairings, and future rosters. From the above instance (Figure 7.1) it is evident that the crew pairing needs to be prepared again such that all the flights within a certain recovery period are covered, and the ToD's outside the recovery period are unchanged.

In the aviation industry, disruption management aspects have been given significant attention; especially in the last few decades. Over this period, numerous techniques and approaches have been applied in aircraft and crew recovery, mostly using heuristics, in various disruptions scenarios to get the disrupted schedule back to normal operations. In the following sections, an overview of the recovery approaches considered by researchers and practitioners is presented.

7.5 AIRCRAFT RECOVERY APPROACHES

During disruptions, ARO gets priority over crew or passenger recovery for reasons such as less complex aircraft scheduling rules, stringent maintenance requirements, fewer disrupted aircraft than disrupted crew and passengers and high aircraft resource value. Decisions involving ARO include flight cancellation, flight delay, aircraft rerouting or a combination of these. Broadly, the solution approaches used in ARO can be classified as network flow algorithms, time–space networks, time–band networks, and connection networks (Clausen et al., 2010). Heuristics and other approaches such as genetic algorithms and grey programming have also been attempted in the ARO.

7.5.1 Solution Approaches Based on Network Flows

In network flow problems, the objects move through the network subject to the capacity constraints of the network. These problems have wide application in solving a range of problems across different industrial settings. Network flow problems are used in the schedule recovery scenarios to reschedule flights and to reroute the aircraft. This approach can also be used to solve the problem of aircraft shortages by delaying or canceling flights. In this, multiple delays and cancellations of flights are allowed, and it permits the use of rerouting, aircraft swapping, and/or using surplus aircraft (Barnhart et al., 1998a). A *switch delay* occurs due to aircraft swap which may have significant impact in addressing schedule recovery in the short haul operations. In this, aircraft maintenance is also taken into consideration to achieve a feasible solution. However, sometimes, different approaches accommodate flight retiming and cancellations but does not take aircraft maintenance into consideration.

Jarrah et al. (1993) present two minimum cost network flow models to reschedule flights and to reroute aircraft when a shortage of aircraft occurs at an airport. The first model is the delay model which solves the problem of aircraft shortages by delaying flights. The objective of the delay model is to minimize the number of rescheduled flights and assignment costs for surplus and recovered aircraft. The second model is a cancellation model in which the objective, in addition to the objective of the delay model, is to capture cancellation costs. Both models allow multiple delays and cancellations of flights and do not prohibit the use of aircraft swapping and surplus aircraft. Neither model captures crew re-pairing and aircraft maintenance into consideration. The models are tested on United Airlines' B737 fleet on three airports.

The delay model proved effective with airports that have a high volume of flights and the cancellation model is considered efficient and applicable with real-time decision support systems.

Based on the model proposed by Jarrah et al. (1993), Rakshit et al. (1996) present a System Operations Advisor (SOA). This real-time decision support system is deployed at United Airlines and has created substantial savings in delay costs as well as reduced potential delays. For a period of six months 692 delay problems were solved and more than 27,000 min of potential delays were saved by SOA which corresponds to $540,000 savings in delay costs.

Mathaisel (1996) presents a schedule disruption scenario in which an aircraft is put out of service due to a maintenance problem. The algorithm presented is based on the network flow algorithm where the objective is to find a feasible flow of aircraft through the disrupted network at minimum cost. The aircraft routing problem is solved for a schedule disruption scenario involving a fleet of six L-1011 aircraft, each with six rotations, covering 19 flights for an international carrier. The model accommodates flight retiming and cancellations but does not take crews and aircraft maintenance into consideration. Also, various business processes and IT challenges are elaborated which are encountered during the implementation of decision support systems for airline disruption management.

Cao and Kanafani (1997a,1997b) present a mathematical model and the computational results separately in parts I and II, respectively. Cao and Kanafani (1997a) present a flight operations decision problem as a quadratic 0–1 programming model to maximize the cost penalties of delays and flight cancellations. This model is based on the delay model of Jarrah et al. (1993), and it considers delay and cancellations simultaneously of the same type of aircraft. An extension of the model is developed to incorporate ferrying of surplus aircraft and the replacement of different types of aircraft. Computational results are provided in the second part of the paper (Cao & Kanafani, 1997b). In this, the computational results are presented for the model presented in part I. The results are presented for 40 experiments involving 20–50 airports, 20–150 aircraft, 5–12 surplus aircraft, and 65–504 flights. The run time for the experiments ranges from 26 to 870 seconds.

7.5.2 Solution Approaches Based on Time–Space Networks

Time–space networks represent location specific scenarios where incoming and outgoing connections are represented against a timeline. In aviation disruption management, various aircraft schedule recovery strategies such as flight cancellations, ferrying of spare aircraft and flight delays are considered. For this, a decision support framework based on several strategies to assist carriers in single or multi-fleet aircraft routings and multi-stop flight scheduling can be used to flow the aircraft in the network at a minimum cost. A framework based on multi-fleet schedule perturbation model can be constructed as a time–space network from which strategic network models are formulated as multi-commodity network flow problems for aircraft schedule recovery. The models are developed on schedule recovery strategies based on flight delays, ferrying of idle aircraft and modification of multi-stop flights.

Yan and Yang (1996) use a time–space network to solve four strategic models based on a basic schedule perturbation model. These models are based on various aircraft schedule recovery strategies such as flight cancellation, ferrying of spare aircraft and flight delays for a single fleet of non-stop flights. The first model considers cancellation of flights and is formulated as a pure network flow problem. The objective of this model is to minimize the costs associated with flights and their cancellations. The second model, also formulated as a pure network flow problem, is an extension of the first model and considers cancellation of flights along with ferrying of a spare aircraft. The third model is a network flow problem with side constraints and considers the delay and cancellation of flights whereas the fourth model incorporates cancellation of flights, ferrying of spare aircraft and flight delays. Breakdown of an aircraft was considered as the cause of schedule perturbation. A case study of a major Taiwan air carrier is presented which involves 15 cities, 12 aircraft, and 319 flights. In all the models, maintenance issues and crew scheduling are not considered which confines the practicality of the models. The models were tested on 252 scenarios. Results show that all pure network flow problems are optimized in less than one minute whereas all network flow problems with side constraints converged to 1% of the optimal solution within 5.5 min.

Yan and Young (1996) present a decision support framework based on several strategic models to assist carriers in multi-fleet routing and multi-stop flight scheduling. These models are formulated as integer multiple commodity network flow problems. In all the models, maintenance and crew issues are not considered. These strategic models are obtained from a basic flight deletion model which is used to evaluate the deletion of uneconomic flights. The basic multi-fleet model is formulated as multiple single-fleet, time–space network problem with the objective to flow all the airplanes in all single-fleet networks at a minimum cost. A case study is taken from the data for a major Taiwan Airline's international operation which includes 24 cities, 26 aircraft, 464 flights of which 367 flights are non-stop while the rest 97 flights are one-stop flights. Three types of aircraft were used in the study type A, B, and C with average seating capacity of 120, 269, and 403, respectively. In all, 25 scenarios were tested with substantial problem sizes.

An extension of the work of Yan and Yang (1996) is presented by Yan and Tu (1997). A framework based on a basic multi-fleet schedule perturbation model (BMSPM) is constructed as a time–space network from which strategic network models are formulated as multi-commodity network flow problems for aircraft schedule recovery. In this, aircraft maintenance and crew scheduling issues were excluded, and temporary shortage of an aircraft is considered as the cause of schedule perturbation. The models were developed on schedule recovery strategies based on flight delays, ferrying of idle aircraft and modification of multi-stop flights. The objective of the strategic models is to minimize the schedule's perturbed period and the loss of profit. A case study based on data from a major Taiwan airline's international operation with 24 cities, 273 flights, and 26 aircraft is conducted. Three types of aircraft were used in the study with average seating capacity of 120, 269, and 403, respectively. Results indicate the models are effective and efficient for handling smaller degrees of schedule perturbations.

The work of Yan and Lin (1997) is also an extension of the work of Yan and Yang (1996) in which a time–space, network-based model was developed from which several perturbed network models were created. The model is developed for operations of a single fleet catering to one-stop and non-stop flights when the schedule is perturbed because of a temporary airport closure. The objective of the basic model is to minimize the perturbed period after an incident and to obtain the most profitable schedule. The strategic models were developed based on schedule recovery options such as flight delays, modification of flights and ferrying of aircraft. Maintenance and crew considerations were not included in the models. A case study based on data from a major Taiwan airline's international operation with 24 cities, 197 flights, and 17 aircraft is conducted. Four types of aircraft were used in the study with average capacity of 269 seats. Ten scenarios were tested on various problem instances. Two types of problems are formulated for these strategic models: pure network flow problems and network flow problem with side constraints. The two pure network flow problems provided solutions in less than 1 s while six network flow problems with side constraints converged within 50 s with 0.1% error.

Thengvall et al. (2000) propose an integer, single commodity network flow model with side constraints for the ARO problem for a single fleet. The proposed model is derived from the work of Yan and his co-authors (Yan & Yang, 1996), (Yan &Young, 1996), and (Yan & Lin, 1997). The model is based on a time–space network with the objective to maximize modified profit associated with the recovery period schedule. The model can accommodate a recovery period of arbitrary length that begins and ends at any time in the day. However, the recovery period used in this paper is one day. The strength of this model is its ability to generate solutions that reflect changing user preferences by adjusting the number of delay operations, the cost of delaying flights, the bonuses awarded for protecting flights, and schedules with different solution properties. Two data sets provided by Continental Airlines are used in the evaluation: for Boeing 757 fleet and for Boeing 737–100 fleet. The Boeing 757 fleet consists of 16 aircraft serving 42 daily flights while the 737–100 fleet consists of 27 aircraft operating at 30 stations and servicing 162 flights daily. The model is tested for the 757 fleet under 108 disruptive situations consisting of all combinations of grounding one, two, and three aircrafts for a maximum delay of 120 min. None of the problem instances took more than one second of run time and overall, 83.2% of all possible flight paths were maintained intact. Twenty of the 30 stations serviced by the 737–100 fleet have one or more aircraft grounded overnight for a maximum length of 90 min and the number of intact flight paths is found to be on average 23 out of 26, and in the worst case 20 of 26. Integer solutions for all 20 cases were found in three seconds. A rounding heuristic (to get feasible integer values) from the LP relaxation was also developed to test eight fractional solutions obtained from runs on Boeing 737–100 fleet. The solutions obtained using the heuristic have more cancellations and intact flight paths on average, but fewer delays and swaps than the integer solutions. The model does not track individual passengers and thus does not consider passenger connections.

Thengvall et al. (2001) present three multi-commodity network flow models for determining a recovery schedule of aircraft following a hub closure. Each proposed model allows cancellations, delays, ferry flights, and substitution between fleets and sub-fleets. Each model is a mixed-integer programming problem. The first model is a preference model which is derived from the work of Yan and his co-authors (Yan & Yang, 1996), (Yan &Young, 1996), (Yan & Lin, 1997), and (Yan & Tu, 1997). This model solves the ARO problem for multiple fleets by maximizing a modified profit over the flight schedule during the recovery period. The objective function accounts for passenger revenues, flight delays, and cancellations. In addition, it includes an incentive to minimize deviations from the original aircraft routings. The second model is the generalized network preference model which is the same as the first one except that it introduces new network modeling constructs that replace the cover constraints in the first model. The third model is an extension (includes multiple fleets) of the model developed by Argüello et al. (1997a) and unlike the first two models is a time band model with the objective to minimize the sum of cancellation and delay costs. In the time band formulation, all station activity is aggregated into discrete time bands. Data was obtained from Continental Airlines for the three hubs Houston, Newark, and Cleveland, spans 2 ½ days and includes 332 active aircraft from 12 different fleets. Each fleet had one to six sub-fleets for a total of 28 different types of aircraft. The schedule includes 2921 flights between 149 domestic and international locations. After comparisons of the solution times and quality of the models, the first model provided high-quality solutions in reasonable time. Maintenance issues were not considered in the assignment of aircraft.

The model presented by Thengvall et al. (2003) was derived from the work of Yan and Tu (1997). The multi-fleet linear programming model allows flight delays and cancellations and accounts for passenger revenues. The proposed model is a time–space model with the objective to maximize the modified profit during the recovery period. Computational results were presented using data from Continental Airlines for 332 aircraft, 12 fleets, 28 different types of aircraft, 2921 flights, 149 domestic and international destinations and three hubs. A total of 27 scenarios were considered, nine for each hub with a recovery period between 240 and 1400 min. All the hubs were closed for a period of two hours. Results indicate that in 16 of the 27 instances the proposed bundle algorithm converged faster than CPLEX's MIP solver. Even in larger instances of 360- and 600-min closing, a feasible solution was found quickly.

7.5.3 Solution Approaches Based on Time–Band Networks

In time–band networks, locations of interest are covered through distinct time intervals. In response to aircraft groundings and delays, Bard et al. (2001) present a model in which the objective of the model was to minimize the flight cancellation costs (profit lost by not operating the scheduled flights) and delay costs (overtime pay and additional fuel) associated with aircraft re-routing. The model was based on the time–band network, introduced by Argüello (1997b). The model was constructed by transforming the problem into a time-based network with discrete time horizon. The resulting model resembled a single commodity minimum cost flow problem with side constraints. Relaxation of the integrality constraints resulted in linear programs which were solved using CPLEX. The algorithms solved the time–band problems efficiently with respect to time and provided a good number of integer solutions. The time-based model was applied to a flight schedule obtained from Continental Airlines for their 737–100 fleet consisting of 162 flights operated by 27 aircraft over a network of 30 stations. A total of 427 instances were tested with five time–band models with 5-, 5/15-, 15-, 15/30-, and 30-min time–band lengths were considered. In particular, the use of 5-min time band led to solutions that were provably within 5% of optimality for 97% of the instances and within 10% of optimality for every instance tested.

7.5.4 Set Partitioning Models Formulated on Connection Networks

In connection networks, nodes represent flights and are connected through arcs. The nodes in the network capture flight information such as origin-destination cities and arrival–departure times. During ARO, heuristics can be used to find a new flight schedule based on the available aircraft which can minimize the total passenger delay on the airline network. Through this, aircraft routings along with a modified flight schedule for a single or multi airline fleet can be obtained. However, other important aspects such as aircraft maintenance, crew rescheduling, and airport curfew generally are not captured. During ARO, in addition to determining new departure times, the approach generates new crew pairings and new aircraft routings.

Clarke et al. (1997) propose a mathematical model that enables simultaneous flight delays and cancellations in a fleet. The objective of the model was to minimize costs associated with reassigning flights to the operational aircraft. These cost coefficients include aircraft direct operating costs, predetermined passenger revenue spill costs, and operating revenue. For this, a greedy heuristic and optimization-based algorithm was developed to solve the Aircraft Schedule Recovery Problem. A case study was conducted on data from a major US domestic carrier to validate and test these algorithms using simulation on various possible constraint scenarios. Results indicate that aircraft utilization in the schedules generated by each algorithm for normal operating conditions is slightly less (within 85%) than that of the actual airline operations. The average aircraft utilization for each scenario under an irregular operating condition is within 95% of that of normal operating conditions.

Rosenberger et al. (2003) propose an optimization model for ARO as a set packing problem that reschedules flight legs and reroutes aircraft with the objective to minimize rerouting and cancellation costs. An aircraft selection heuristic was developed for selecting which aircraft are rerouted and provide proof of concept by evaluating the model using a simulation for airline operations. The approach is based on two assumptions. First, airlines assign legs to a given set of arrival slots and second, ARO reroutes aircraft only from the same fleet. ARO was validated using a simulation (SimAir) on a daily flight schedule of on fleets from a major domestic airline in the United States. Three fleet types were considered with 198 aircraft corresponding to 910 flights in total. A delay threshold of 30 min (to use ARO) with a simulation time of 500 days of flight operations was considered. ARO allowed delays of up to three hours in the construction of new routes to minimize the number of cancellations. As compared to the shortest cycle cancellation policy, for each fleet, ARO significantly reduces cancellations and passenger disruptions, while improving on-time performance.

Andersson and Värbrand (2004) develop a mixed integer multi-commodity flow model with side constraints allowing cancellations, delays, and flight swaps between different aircraft types. The objective of the model was to pick one route for each aircraft so that the total revenue is maximized. The model was reformulated in a set packing model using Dantzig–Wolfe decomposition, where each column represents a feasible route for a certain aircraft. Two column generation schemes were used to find heuristic solutions quickly. The models were tested on two data sets obtained from a Swedish domestic airline. The first data set considers 13 aircraft, 2 aircraft types, 98 flights, and 19 airports while the second had 30 aircraft, 5 aircraft types, 215 flights, and 32 airports. Flight durations were short ranging between 15 and 125 min with turnaround time of 10 min. Issues such as maintenance considerations, crew constraints, and slot times were not considered.

7.5.5 Heuristic Approaches

Heuristics approaches are based on practical problem-solving methods that do not necessarily conform to algorithms. Heuristics-based results do not guarantee optimal solutions and are preferred in resource constrained scenarios or in conditions when a solution is impractical to obtain due to lack of information.

Teodorović and Guberinić (1984) made one of the first attempts to address the airline schedule perturbation problem where aircraft were taken out of service due to technical difficulties. To find a new flight schedule, based on the available aircraft, a heuristic was proposed which minimizes the total passenger delay on an airline network. A branch and bound algorithm was used to determine aircraft routings along with a modified flight schedule for a single airline fleet. An example involving eight flights on a fleet of three aircraft with 25–110 passengers was investigated with the assumption that all aircraft have the same capacity. In this, operational daily airline scheduling problem issues like the number of passengers for each flight are considered but not other important aspects such as aircraft maintenance, crew rescheduling, and airport curfew.

Teodorović and Stojković (1990) extended the model proposed by Teodorivic and Guberinić (1984). In this, situation was considered where one or more aircraft were canceled. A heuristic algorithm based on dynamic programming was proposed to determine new flight departure times and routings for the operational aircraft in the network. The algorithm was based on a sequential approach where aircraft are processed in sequence. The model had two objective functions. The higher priority one minimizes the number of canceled flights and the lower priority one minimizes the total time loss of passengers on the flights that are not canceled. A fleet of 14 aircraft of the same capacity with 80 flights were considered for computational purposes. A disruption instance at the beginning of the day (with one canceled aircraft) was tested. The average run time for an instance with 12–80 flights were found to be between 5.5 and 140 s. In the model, flight cancellation and airport curfews were taken into account whereas crew scheduling and ferrying of aircraft was not considered.

Teodorović and Stojković (1995) extend their previous work (Teodorović & Stojković, 1990) by including crew considerations in the model. A heuristic approach based on dynamic programming was presented to solve the model. The first objective, to maximize the total number of operating flights, was given priority over the second objective function i.e., the total passenger time loss on flights that were not canceled. In addition to determining new departure times, the model generated new crew pairings and new aircraft rotations. The model was tested on 240 different numerical examples each with four to five arbitrarily generated disturbances with a total of over 1000 different situations. The problems with 80 flight legs using First-In-First-Out principle and a sequential approach using dynamic programming were solved in 2 and 140 s, respectively. Maintenance issues and airport curfew were considered in the model, but computational results provided were inadequate to test the feasibility and applicability of the model in real situations.

Clarke et al. (1996) presented an integer programming model incorporating yield management, vehicle routing, maintenance scheduling, and crew scheduling considerations. During irregular operations, the model addressed the problem of aircraft assignment for all operational aircraft in the fleet by minimizing cost associated with reassigning flight sequences. However, the model was not tested in real-life situations. The solution procedure was developed around the framework of a three-phase decision process. First, to *generate* potential aircraft rotations using modified tree search algorithms. Second, to *assign* aircraft rotations to each operating aircraft while optimizing (maximizing profit or minimizing cost). Third, to *revise* the overall network structure by adjusting scheduled arrival and departure times of each flight.

A swapping heuristic was presented by Talluri (1996) satisfying the basic requirements of a daily fleet assignment model such as flow balance, flight coverage, and equipment count. Two algorithms with polynomial running time were presented for creating swapping opportunities for overnight equipment type changes. The first algorithm finds a swap opportunity that does not involve changing any of the overnight equipment types whereas the second algorithm allows at most k overnight changes if the first algorithm fails to get a solution. Swaps between two equipment types were restricted at any given time. Results indicate that the algorithm, for same-day swap opportunities, provides an improvement over the heuristic proposed by Berge and Hopperstad (1993) in two ways. One, it takes less time and second, it guarantees a swap, in case of a same-day swap opportunity which is helpful in the fast recovery of a schedule. Maintenance and crew constraints were not considered, and identical ground times were always assumed for all aircraft.

In response to grounding and delays, Argüello et al. (1997b) developed a greedy randomized adaptive search procedure (GRASP). The objective of the integer programming model was to minimize the flight cancellation and delay costs associated with ARO. Data were taken from Continental Airlines' 757 fleet for 42 flights, 16 aircraft, and 13 airports. In total, 6068 problem instances were generated by grounding 1–5 aircraft at the beginning of the day for a recovery period of one day. Each aircraft fleet was considered separately so that within each problem instance all aircraft are interchangeable. Maintenance issues along with crew constraints were not considered in the model. Results show that in more than 70% of all instances, the GRASP solution was within 5% of optimality.

Løve et al. (2005) proposed a steepest ascent local search (SALS) heuristic to solve the ARO problem. The heuristic was tested on the data of ten flight schedules from British Airways (BA). Each flight schedule consisted of all the BA short haul flights. Realistic delays (between 30 and 210 min) were introduced on approximately 20% of all aircraft. The average size of the instances was 79 aircraft, 44 airports, 339 flights, and 16 disrupted lines of work. The results indicated an improvement in the processing time of the recovery problem. On average, feasible revised schedules were obtained with in less than 10 s for all planned flights making SALS attractive.

Anderson and Granberg (2006) proposed two meta-heuristics (Tabu Search and Simulated Annealing) both utilizing path relinking (PR) which were developed and tested for the Flight Perturbation Problem (FPP) in which ARO was carried out followed by crew and passenger recovery. In PR, two solutions: initial and guiding were selected from the set of 'elite' solutions that were found during the search. From the initial solution to the guiding solution, a path of solutions was generated. Considering an aircraft breakdown or delays on several flights in the set, the heuristics were applied to three data sets with 13–58 aircraft, 2–5 aircraft types, 98–436 flights, and 19–41 airports on a Swedish domestic airline. The simulated annealing heuristic does produce acceptable solutions, but in general seems less successful than the Tabu search heuristic. Tabu search always produced a solution, within 15 s, that is less than 0.3% from the best-known solution.

7.5.6 Other Approaches

Luo and Yu (1997) developed an integer programming model to overcome schedule perturbation caused by a ground delay program with the objective to minimize the total number of delayed flights. Data were taken from American Airlines which involved 71 incoming flights, 70 of which are delayed by more than 15 min. Similarly, 42 outgoing flights were considered among which 22 require both crew and aircraft from incoming flights. The model was solved in 16.2 s. A heuristic was presented to find feasible solutions and results showed that realistic problems were solved to optimality within seconds.

Stojković et al. (2002) attempted to restore a planned airline schedule without changes to aircraft itineraries and crew pairings. The proposed model re-optimized departure times to consider the sequences of activities that were to be carried out within all aircraft routes and crew pairings with the objective to minimize the total cost. Also, the dual of the model was presented as a network problem. Ten test problems with different dimensions were generated ranging from 2–900 aircraft itineraries, 5–2250 crew rotations, 12–5400 flights, 3–1350 crew connections, and 1–450 passenger connections. The primal was solved by the simplex method whereas the dual of the problem was solved using the network simplex algorithm. The run time for the primal ranged from 0.02 to 29.60 s, while the dual times ranged from 0.02 to 6.71 s. Overall, execution times for the primal increased with the increase in the problem size.

Xiuli and Jinfu (2007) proposed a grey programming method to solve the Flight Schedule Recovery Problem. A grey system contains information presented as grey numbers (random variables) and a grey decision is a decision made within a grey system. The model was a transformed model of Jarrah et al. (1993) solved using a heuristic. The model solved the problem of aircraft shortages at a station by delaying flights until the aircraft shortage is recovered. The objective of the model was to minimize the costs incurred in rescheduling all the flights, in assigning flights to recovered aircraft and in assigning flights to surplus aircraft. The costs include delay and/or swap costs in addition to ferrying costs for the surplus aircraft.

7.6 CREW RECOVERY APPROACHES

In disruption management, as compared to ARO, crew recovery is a complex issue. Most aspects on the crew recovery problem are tackled once the flight schedule is fixed or modified during disruptions. The objective of the crew recovery problem is to re-pair the broken pairings in such a way that crew scheduling rules are satisfied during operations. Some of the common ways to obtain crew recovery is to delay, cancel, swap the flights, utilize reserve crew and re-position crew by flying them as passengers (Barnhart & Shenoi, 1998b). A *reserve crew* is a crew on call, staying at home ready to work if required. A reserve crew has minimum guaranteed hours paid even if no duty is performed, which makes it as an expensive resource. Some airlines have 25–30% additional personnel in the form of reserve crew (Sohoni et al., 2006). Crew recovery approaches can be classified with respect to fixed flight schedule, flight cancellations, and departure delays (Clausen et al., 2010).

7.6.1 Crew Recovery with Fixed Flight Schedule

Crew recovery with fixed flight schedule can be considered as an integer multi-commodity network flow problem where each crew member (including reserve crew) represents a commodity can be used to minimize the total number of modified pairings by covering the maximum number of flights. First, the flight schedule is considered and fixed and is given to re-pair the broken pairings in the modified schedule. The objective in crew recovery is to minimize the cost of pairings along with costs associated with covering a set of tasks. During crew recovery, it is assumed that the ARO problem is solved, and the aim of the problem is to provide a crew for each flight in the modified schedule. This approach integrates crew pairings and crew rostering together.

Wei and Yu (1997) propose a search heuristic for crew re-pairing problem in the case of minor flight perturbations. An integer multi-commodity network flow problem where each crew member (including reserve crew) represents a commodity is presented with the objective to minimize the total number of modified pairings by covering the maximum number of flights. It is assumed that the flight schedule has been fixed and is given to re-pair the broken pairings in the modified schedule. The model is good for a single equipment type and captures all the important elements of the crew repair problem: ports, crews, and flights. A 'generate-and-test' approach is adopted to solve the problem in which one or a few pairings are modified or generated to test the status of the problem. The problem, at each node of the search tree, is defined by the set of uncovered flights and a list of modified pairings. Computational results are shown for two problem instances. Instances with 18 flights of the DC 9 fleet have six pairings and one reserve crew for a same-day recovery period. Six test cases were generated for this problem instance using crew swapping, flight cancellation, and different flight delay scenarios to test the algorithm for the problems. All the cases were solved between 0.72 and 24.22 s for one or more solutions. The other problem instance was tested for 6 airports, 51 flights, and 18 pairings. In this problem instance, eight cases were tested and for most of them three solutions were obtained within a solution time of 0.29 to 5.76 s. Few computational results were presented for this problem instance.

Stojković et al. (1998) presented an optimization approach that solved the problem of crew pairing and crew rostering in a modified flight schedule using column generation approach. The data were taken from a US carrier and only cockpit personnel (pilots), of a specific aircraft type, who are positioned at a particular base, was considered. The disturbances considered in the test problems originate from three delayed flights and one indisposed crew member. Two problems were generated from the input data where both the problems had a working period of 13 days. These problems were further divided into two test scenarios. The first scenario in both problems had an operational period of one day while the second scenario had an operational period of seven days. A total of 106 and 210 flight segments were considered for the first and second problem, respectively. All the test problems were solved in reasonable times, ranging from a few seconds to 20 min.

Guo et al. (2005) propose a genetic algorithm-based approach to solve the aircraft crew recovery problem. It is assumed that the ARO problem was solved, and the aim of the problem was to provide a crew for each flight in the modified schedule. This approach integrated crew pairings and crew rostering. The proposed model was a set partitioning model with the objective to minimize the total operational cost and the cost associated with crew disturbances. Computational results were obtained from a European airline which operates in a point-to-point type of network. Three disruption scenarios were considered: major, medium, and minor. The number of affected flights and affected crew members for minor, medium, and major disruptions were 1–2, 4 and 5–13, with a recovery period of 15–20, 24–40, and 72–111 h, respectively. Results show that most minor and medium instances were feasible but for larger problems it was difficult to find optimal solutions.

A duty-period-based network model was presented by Nissen and Hasse (2006). The model was formulated for a major European airline which minimizes the rescheduling cost during schedule disturbances. A flight schedule which complies with the German regulatory framework was used to generate two crew schedules: one for short-haul and another for medium-haul flights. The framework used a hub-and-spoke structure with a single hub and for a period of one week. The short-haul schedule had eight routes and covered 450 flights from one fleet, whereas the medium-haul schedule had 35 routes and covered 927 flights also from one fleet. For both schedules 14 scenarios were considered and for each of them five test cases were generated. A recovery horizon of 12, 24, 48, 60, and 96 h was chosen for observation. Results for the short-haul and medium-haul schedules show a recovery period of 48 h and 60 h, respectively, producing the best compromise between solution quality and time.

Medard and Sawhney (2007) attempted to integrate crew pairing and rostering models by applying a generate and optimize technique in a single step to solve the crew recovery problem. The integer programming model is the flight-based equivalent to the original pairings based rostering model where the flights have replaced the pairings. The model incorporated the pairing and rostering dimensions of the recovery problem such as government regulations, minimum rest time, total flight/duty time, and rosters. A total of 19 instances were tested of which nine were for a single base, seven were for multi-base and three for roster maintenance. Results show the instances for a single base and multi-base problems for a recovery time window of 48 h. On small to medium size instances (up to 500 crew members), the solution quality improved. However, on larger multi-base problems, too much computational time was required (up to 63% of the total running time) to set up the duty network.

7.6.2 Crew Recovery with Flight Cancellations

Flight cancellations leads to crew recovery in three instances of irregular operations (unplanned maintenance, reduced landing capacity, and airport shutdown) with varying levels (small, medium, and large) of disruptions. This approach first captures the uncovered flights and repairs the broken pairings. Then, new pairings are generated after the assignment of a crew to an uncovered flight and lastly, legality of these pairings is checked.

Lettovský et al. (2000) presented an optimization-based solution approach to solve the crew recovery problem in a disrupted schedule. A crew recovery model (CRM) was presented as a set covering problem to find the minimum cost set of pairings that cover as many flights as possible with minimum impact on passengers (in case of flight cancellations and delays). The model was used for small to medium size disruptions. The schedule consisted of 1296 flight legs that represented 177 pairings originating from two crew bases of a major US airline. Three possible scenarios of irregular operations (unplanned maintenance, reduced landing capacity, and airport shutdown) with different levels (small, medium, and large) of disruptions were presented. The computational results demonstrated that medium-sized disruptions can be handled within an acceptable running time of 2–115 seconds while having 7–2 uncovered flights. Large disruptions were solved with a solution time of 6–97 s having 21–6 uncovered flights whereas small disruptions were solved in 1–6 s with no uncovered flight.

The model proposed by Yu et al. (2003) was similar to the model proposed by Lettovský et al. (2000). However, a heuristic-based search algorithm was proposed with a generate-and-test approach for a real-time decision support system called CrewSolver. The model presented for crew recovery is based on a set covering model with the objective to minimize the costs associated with pairings, uncovered flights, deadheading, and unpaired crew. The proposed algorithm first captured the uncovered flights and repaired the broken pairings. New pairings were generated after the assignment of a crew to an uncovered flight and lastly, legality of these pairings was checked. Operational data were taken from one fleet type of Continental Airlines for 11,847 pairings, 12,390 crew, and 43,625 flights. Computational results were presented for different categories for 1–40 affected flights, 17–243 disruption scenarios, and an average solution time ranging between 0.97 and 321.13 s. Continental Airlines estimated that it saved approximately $40 million to recover from four major disruptions in 2001.

7.6.3 Crew Recovery with Departure Delays

Advanced optimization approaches can solve the crew and flight scheduling problems simultaneously as compared to the traditional sequential approach where the flight schedule needs to be fixed first and thus given as input data to the crew scheduling problem. This proactively recovers projected crew problems which arise due to disruptions before their occurrence by using reserve crew, crew swapping (*Move-up* crews), standby (*reserve*) crew, and deadheading.

Stojković and Soumis (2001) presented an integer nonlinear multi-commodity network flow model with time windows and additional constraints for the operational pilot scheduling problem. The objective of the model was to minimize the difference between the new and the planned departure times, and to minimize the total number of pilots whose next-day duties are influenced by modifications to a given day of operations. This work was similar to the previous work of Stojković et al. (1998) to find simultaneous recovery of crew parings and crew rostering. The model was implemented and tested on three data sets of domestic flights by assuming a hypothetical disturbance that caused the closure of the hub airport for one hour. Three problem sets relate to a perturbation of a small, medium, and large fleet. Problem 1 with 18 pilots, 28 aircrafts, and 66 flights correspond to small fleet while problem 2 and problem 3 with 42 and 59 pilots, 58 and 79 aircraft and 131 and 190 flights represent medium and large fleets, respectively. Results for three input data sets show that the proposed approach is, on average, longer than the traditional one but solution quality is found to be better than the traditional approach as the number of uncovered flights is significantly decreased.

Abdelghany et al. (2004) present a heuristic-based decision support tool that automates crew recovery for hub-and-spoke networks. The proposed tool proactively recovered projected crew problems, arising due to disruptions before their occurrence, by using reserve crew, crew swapping, standby crew, and deadheading. The tool adopted a rolling approach in which a sequence of optimization assignment problems was solved such that it recovered flights in chronological order of their departure times. The objective of each assignment problem was to recover as many flights as possible while minimizing total system cost resulting from resource reassignments and flight delays. A total of 18 crew disruption problems for one day of operations, seven of which were misconnections and two of which were duty problems, were considered. The remaining nine were rest problems for one day of irregular operations. All instances relate to the cockpit crew (captain and first officer) for four different fleets. A total of 121 crew members were considered which include 18 stranded, 17 standbys, 52 reserve, and 34 undisrupted pilots. Computational results show that open positions (4 flights), 8 deadheads, 4 standbys, 1 reserve, and 6 swaps among the pilots were observed to solve the problem in less than 2 min. One of the limitations of the model was that it does not take aircraft and cabin crew into consideration.

7.7 PASSENGER AND INTEGRATED RECOVERY

Passenger recovery results after a disruption in a schedule and is desired by the airlines to re-pair the itineraries of disrupted passengers. Passenger recovery is carried out after aircraft and crew recovery. The problem is decomposed into passenger review (determine which passengers are impacted by disruption), passenger aggregation (combine similar disrupted passengers), itinerary generation (identify shortest permitted path), and seat allocation (determine optimum passenger allocation).

Ground transportation is considered as an alternative for passenger recovery by air, subject to ground transportation times. However, integrated recovery of aircraft, crew, and passengers is a difficult task. Conflicting objectives of the important airline resources i.e. aircraft and crew make it difficult to integrate them as one problem which yield an optimum solution that is optimal for the integrated problem. Few attempts have been made to integrate the recovery of resources that provide good feasible solutions, if not optimal, during the recovery process.

Clark (2005) proposed a linear programming model for disrupted passenger recovery. The aim of the model was to determine an optimum recovery solution for passengers and the airline. The model was solved as a multi-commodity network flow problem over a time–space network with the objective to minimize the incremental costs to the airline while maximizing passenger satisfaction. The model was tested for Air New Zealand data, and it was found that it is highly dependent on external systems for passenger information and distribution.

Bratu and Barnhart (2006) presented two models to simultaneously find recovery plans for aircraft, crew, and passengers. The first model, called Disrupted Passenger Metric (DPM) minimized operating and disrupted passenger costs whereas the second model, Passenger Delay Metric (PDM) minimized operating and passenger delay costs. In both models, aircraft maintenance requirements were not considered. Instead, aircraft routings were generated that positioned maintenance critical aircraft at specific maintenance positions. An airline operations control center simulator was developed to test the models on the data of a major domestic US airline. Data were comprised of four different fleet types with 302 aircraft, 83,869 passengers on 9,925 different passenger itineraries, and 74 airports with three hub airports. More than two-thirds of the passengers had single-leg itineraries while the rest were connecting through one or more of the hubs. The recovery models were tested for first, second, and third day of operations corresponding to high, average and low levels of disruption, respectively. On the first day of operations a total of 1063 flights, of which 61.1% are delayed and 60 are canceled, were tested with both PDM and DPM. Results for this instance show that DPM provides recovery solutions in real time whereas PDM requires excessive solution time. The second day of operations had 1061 flights where 53.2% were delayed and 29 were canceled. As compared to actual operations, DPM-reduced passenger delay (in minutes) by 8.2%, the number of disrupted passengers by 19.4% and cancellations by 8.5%. Similarly, for the third day of operations which had 1074 flights of which 34% are delayed and 9 are canceled, DPM provided better solution as compared to actual operations with respect to passenger delay (in minutes) by 5.2%, number of disrupted passengers by 12.3% and cancellations by 2.6%. Overall, it was found that PDM cannot be solved in real time for day-long decision windows and days with relatively high levels of disruptions, whereas DPM was fast enough to recover from airline irregularities in real time.

Zhang and Hansen (2008) considered passenger recovery from schedule perturbations caused by inclement weather or temporary closure of hub airports and proposed a real-time intermodal substitution (RTMIMS) strategy during disruptions. Ground transportation was considered as an alternative for passenger recovery by air, respecting ground transportation times. The objective of the model was to minimize the passenger costs due to delay or cancellation as well as to minimize the operating costs of the transportation. An example with 40 flights with a four-hour time window was considered. Results show that intermodal transportation substitution reduced the number of disrupted passengers from 90 to 14.

Similarly, integrated recovery of aircraft, crew, and passengers is a difficult task. Conflicting objectives of the important airline resources i.e. aircraft and crew make it difficult to integrate them as one problem which yield an optimum solution that is optimal for the integrated problem. Few attempts have been made to integrate the recovery of resources which provide good feasible solutions, if not optimal, during the recovery process. Lettovský (1997) attempted to integrate aircraft, crew, and passengers. The proposed Schedule Recovery Model (SRM) maximizes total profit to the airline while capturing all the three mentioned entities. The SRM determines a plan for equipment assignment, cancellation, and delays considering landing restrictions. For each equipment type, an ARO model was solved and for each crew group the CRM is solved before the passenger evaluation model was evaluated.

Abdelghany et al. (2008) presented an integrated Decision Support Tool for Airlines schedule Recovery (DSTAR) during irregular operations. The tool implemented a greedy optimization approach to integrate a schedule simulation model and a resource assignment optimization model. The model was an extension of the CRM of Abdelghany et al. (2004). The objective of the model was to minimize the resource assignment cost, total delay cost and cancellation cost. Deadheading was not considered in the recovery plan. An application of DSTAR was tested on a major US air carrier during a hypothetical Ground Delay Program (GDP) scenario with a maximum delay of 110 min for 522 aircrafts, 1360 pilots, 2040 flight attendants, and 1100 daily flights serving 112 cities. The impact of the GDP using the flight simulation model resulted in 177 disrupted flights with 7633 total delay minutes. The recovery process was divided into nine independent stages, and all are generated in less than 40 s.

7.8 CONCLUSION

Over time, the upsurge in air traffic has resulted in airlines choosing to highly optimize the utilization of their resources, to the extent of leaving smaller time windows and/or idle resources to deal with unplanned operations. A disrupted schedule causes a cascade of disturbances across the flight network, further affecting resources like aircraft and crews. Following a disruption, decisions involving ARO are always given priority over both, crew and passenger recovery. After the flight schedule has been fixed, crew recovery problems are addressed by delaying, canceling, swapping as well as utilizing the reserve crew. For the purposes of schedule recovery, a range of optimization models and heuristic approaches have been adopted by researchers and experts, helping airlines save millions of dollars each year.

CHAPTER QUESTIONS

Q1. Discuss the concept of disruption management in the aviation operations. Also, identify and comment on all possible disruption scenarios.
Q2. Discuss three major criteria considered during schedule recovery operations. Justify the selection of each criterion with examples.
Q3. How is ARO achieved during disruptions? Comment on three ARO approaches.
Q4. How is crew recovery achieved during disruptions? Comment on three crew recovery approaches.
Q5. Discuss the two advantages and disadvantages of integrated recovery approach. How does it affect schedule recovery?

REFERENCES

Abdelghany, A., Ekollu, G., Narasimhan, R., & Abdelghany, K. (2004). A proactive crew recovery decision support tool for commercial airlines during irregular operations. *Annals of Operations Research, 127*(1), 309–331.

Abdelghany, K. F., Abdelghany, A. F., & Ekollu, G. (2008). An integrated decision support tool for airlines schedule recovery during irregular operations. *European Journal of Operational Research, 185*(2), 825–848.

AhmadBeygi, S., Cohn, A., & Lapp, M. (2010). Decreasing airline delay propagation by re-allocating scheduled slack. *IIE Transactions, 42*, 478–489.Anderson, T., & Granberg, T. A. (2006). Solving the flight perturbation problem with meta heuristics. *Journal of Heuristics, 12*(1), 37–53.

Andersson, T., & Värbrand, P. (2004). The flight perturbation problem. *Transportation Planning and Technology, 27*(2), 91–118.

Argüello, M. F., Bard, J. F., & Yu, G. (1997a). Models and methods for managing airline irregular operations. In Yu, G. (Ed.), *Operations Research in the Airline Industry* (pp. 1–45). Kluwer Academic Publishers.

Argüello, M. F., Bard, J. F., & Yu, G. (1997b). A GRASP for aircraft routing in response to groundings and delays. *Journal of Combinatorial Optimization, 1*(3), 211–228.

Ball, M., Barnhart, C., Nemhauser, G., & Odoni, A. (2007). Air transportation: Irregular operations and control. In C. Barnhart, & G. Laporte (Eds), *Handbook in OR & MS* (pp. 1–167). Elsevier.

Bard, J. F., Yu, G., & Argüello, M. F. (2001). Optimizing aircraft routings in response to groundings and delays. *IIE Transactions, 33*(10), 931–947.

Barnhart, C., Boland, N. L., Clarke, L. W., Johnson, E. L., Nemhauser, G. L., & Shenoi, R.G. (1998a). Flight string models for aircraft fleeting and routing. *Transportation Science, 32*(3), 208–220.

Barnhart, C., & Shenoi, R.G. (1998b). An approximate model and solution approach for the long-haul crew pairing problem. *Transportation Science, 32*(3), 221–231.

Berge, M. E., & Hopperstad, C. A. (1993). Demand driven dispatch: A method for dynamic aircraft capacity assignment, models and algorithms. *Operations Research, 41*(1), 153–168.

Bratu, S., & Barnhart, C. (2006). Flight operations recovery: New approaches considering passenger recovery. *Journal of Scheduling, 9*, 279–298.

Cao, J. M., & Kanafani, A. (1997a). Real-time decision support for integration of airline flight cancellations and delays part I: Mathematical formulation. *Transportation Planning and Technology*, 20, 183–199.

Cao, J. M., & Kanafani, A. (1997b). Real-time decision support for integration of airline flight cancellations and delays part II: Algorithm and computational experiments. *Transportation Planning and Technology*, 20, 201–217.Clarke, L., Johnson, E., Nemhauser, G., & Zhu, Z. (1997). The aircraft rotation problem. *Annals of Operations Research*, *69*, 33–46.

Clarke, L.W., Hane, C. A., Johnson, E. L., & Nemhauser, G. L. (1996). Maintenance and crew considerations in fleet assignment. *Transportation Science*, *30*(3), 249–260.

Clark, M. D. D. (2005). Passenger re-accommodation a higher level of customer service, AGIFORS Airline Operations Study Group Meeting, Germany.

Clausen, J., Larsen, A., Larsen, A., & Rezanova, N. J. (2010). Disruption management in the airline industry—concepts, models and methods. *Computers and Operations Research*, *37*(5), 809–821.

Guo, Y., Suhl, L., & Thiel M. P. (2005). Solving the airline crew recovery problem by a genetic algorithm with local improvement. *Operational Research: An International Journal*, *5*(2), 241–259.

Jarrah, A. I., Yu, G., Krishnamurthy, N., & Rakshit, A. (1993). A decision support framework for airline flight cancellations and delays. *Transportation Science, 27*, 266–280.

Lettovský, L., (1997). Airline operations recovery: An optimization approach, PhD Thesis, Georgia Institute of Technology: Atlanta, GA.

Lettovský, L., Johnson, E. L., & Nemhauser, G. L. (2000). Airline crew recovery. *Transportation Science*, *34*(4), 337–348.

Løve, M., Sørensen, K. R., Larsen, J. & Clausen, J. (2005). Using heuristics to solve the dedicated aircraft recovery problem. *Central European Journal of Operations Research*, *13*(2), 189–207.

Luo, S., & Yu, G. (1997). On the airline schedule perturbation problem caused by the ground delay program. *Transportation Science*, *31*(4), 298–311.

Mathaisel, D. F. X. (1996). Decision support for airline system operations control and irregular operations. *Computers & Operations Research*, *23*, 1083–1098.

Medard, C. P., & Sawhney, N. (2007). Airline crew scheduling from planning to operations. *European Journal of Operations Research*, *183*, 1013–1027.

Nissen, R., & Hasse, K. (2006). Duty-period-based network model for crew rescheduling in European airlines. *Journal of Scheduling, 9*, 255–278.

Rakshit, A., Krishnamurthy, N., & Yu, G. (1996). System operations advisor: A real-time decision support system for managing airline operations at United airlines. *Interfaces*, *26*, 50–58.

Rosenberger, J. M., Johnson, E. L., & Nemhauser, G. L. (2003). Rerouting aircraft for airline recovery. *Transportation Science*, *37*, 408–422.

Sohoni, M., G., Johnson, E. L., & Bailey, T. G. (2006). Operational airline reserve crew planning. *Journal of Scheduling*, *9*, 203–221.

Stojković, M., & Soumis, F. (2001). An optimization model for the simultaneous operational flight and pilot scheduling problem. *Management Science*, *47*(9), 1290–1305.

Stojković, M., Soumis, F., & Desrosiers, J. (1998). The operational airline crew scheduling problem. *Transportation Science, 32*, 232–245.

Stojković, G., Soumis, F., Desrosiers, J., & Solomon, M. M. (2002). An optimization model for a real-time flight scheduling problem. *Transportation Research Part A, 36*, 779–788.

Talluri, K. T. (1996). Swapping applications in a daily airline fleet assignment. *Transportation Science*, *30*, 237–248.

Teodorović, D., & Guberinić, S. (1984). Optimal dispatching strategy on an airline network after a schedule perturbation. *European Journal of Operations Research, 15*, 178–182.

Teodorović, D., & Stojković, G. (1990). Model for operational daily airline scheduling. *Transportation Planning and Technology, 14*, 273–285.

Teodorović, D., & Stojković, G. (1995). Model to reduce airline schedule disturbances. *Journal of Transportation Engineering, 121*, 324–331.

Thengvall, B. G., Bard, J. F., & Yu, G. (2000). Balancing user preferences for aircraft schedule recovery during irregular operations. *IIE Transactions, 32*, 181–193.

Thengvall, B. G., Bard J. F., & Yu, G. (2003). A bundle algorithm approach for the aircraft schedule recovery problem during hub closures. *Transportation Science, 37*(4), 392–407.

Thengvall, B. G., Yu, G., & Bard, J. F. (2001). Multiple fleet aircraft schedule recovery following hub closures. *Transportation Research Part A, 35*, 289–308.

Wei, G., & Yu, G. (1997). Optimization model and algorithm for crew management during airline irregular operations. *Journal of Combinatorial Optimization, 1*, 305–321.

Xiuli, Z., & Jinfu, Z. (2007). Grey programming for irregular flight scheduling. In Proceedings of the IEEE International Conference on Grey Systems and Intelligent Services, 164–168.

Yan, S., & Lin, C. (1997). Airline scheduling for the temporary closure of airports. *Transportation Science, 31*, 72–82.

Yan, S., & Tu, Y. P. (1997). Multi-fleet routing and multi-stop flight scheduling for schedule perturbation. *European Journal of Operational Research, 103*, 155–169.

Yan, S., & Yang, D. H. (1996). A decision support framework for handling schedule perturbation. *Transportation Research Part-B, 30*(6), 405–419.

Yan, S., & Young, H. (1996). A decision support framework for multi-fleet routing and multi-stop flight scheduling. *Transportation Research Part A: Policy and Planning, 30*(5), 379–398.

Yu, G., Argüello, M., Song, G., McCowan, S., & White, A. (2003). A new era for crew recovery at Continental Airlines. *Interfaces, 33*(1), 5–22.

Zhang, Y., & Hansen, M. (2008). Real–time intermodal substitution strategy for airline recovery from schedule perturbation and for mitigation of airport congestion. *Transportation Research Records, 2052*(1), 90–99.

8 Airline Disruption Management – II

CHAPTER OBJECTIVES

At the end of this chapter, you will be able to

- Understand the issues and considerations related with airline schedule recovery.
- Understand the concept of disruption neighborhood.
- Create a network of disrupted resources and tasks.
- Understand the concept of column generation.
- Understand the set partitioning model formulation.
- Know the rolling time horizon recovery approach.
- Know about multi-objective and multi-fleet schedule recovery aspects.

8.1 INTRODUCTION

Disruptions are a rather frequent occurrence in airline operations. They are considered part and parcel of airline operations and a consistent problem area that must be managed in the most efficient and effective of ways. This chapter aims to explore disruption management within the purview of two key airline resources, i.e., aircraft and crew. On the day of operations schedule disruptions lead to logistical problems for both the airline as well as the passengers. Once the schedule disruption is identified, on the basis of different recovery approaches, airlines attempt to get back to the originally planned schedule as soon as possible. As appropriate, illustrations are presented in this chapter to elaborate on a range of airline disruption management scenarios and phases.

8.2 THE CONCEPT OF DISRUPTION NEIGHBORHOOD

Neighborhood search approaches are considered for airline schedule recovery (Guimarans et al., 2015; Ng et al., 2020). In this chapter, schedule recovery in airline operations is considered by using the concept of disruption neighborhood. This concept is presented by Rezanova and Ryan (2010) to address the train driver recovery problem for Danish passenger railway operator DSB S-tog A/S. For details on railway disruption management, refer to the supplementary chapter. Regarding the

DOI: 10.1201/9780203731338-8

application of disruption neighborhood approach in the aviation industry, the concept of neighborhood generation is similar to the approach used in the rail industry.

In the schedule, a flight is assigned an aircraft, a captain, and a first officer – the most important resources to operate a flight. To recover a disrupted schedule, a disruption neighborhood of the affected flights and resources are generated in which new tasks may be assigned to resource(s) which are included within the neighborhood. A task is an allocation of activities a resource would do if it is included in the neighborhood; it may be a new task or an existing one in the schedule. Tasks vary from resource to resource. For example, aircraft may be given a *flight task* or a *maintenance task*, whereas for crews a task could be a *flight task* or *deadheading* or a *training task*.

During the generation of neighborhood, delaying of flight(s) and/or swapping of resources is also considered. For instance, a captain may replace a first officer so that two captains cover a flight in the neighborhood. However, two first officers are not assigned together to cover a flight. In the disruption neighborhood, these resources are given new tasks to perform which are only valid within the neighborhood. Furthermore, making use of the idle resources in the neighborhood is also considered to get back to the original schedule. Idle resources are the resource(s) which have arrived at the port of disruption before the time of disruption but are scheduled to depart after the initial recovery time window ends. For example, an aircraft which has more ground time than the required minimum turn-around time or a reserve crew on ground are considered as idle resources. Similarly, crew can also perform any task(s) if their recovery duties are feasible in the neighborhood. During schedule recovery operations, one of the major objectives is to minimize the total cost of feasible duties for each resource included within the disruption neighborhood.

In this idea of disruption neighborhood to tackle airline schedule recovery problems, initially, a small set of disrupted resources and flights are considered to recover the schedule within a certain recovery period. If feasible solution is not achieved, the disruption neighborhood is expanded until all resources are assigned a feasible task to perform in the neighborhood. Once the resources come out of the neighborhood, they resume their originally planned duties for the remaining day's operations as planned in the schedule.

Generation of resources and tasks for disruption neighborhood is dependent on the magnitude of the disruption. If the disruption is severe, then the number of disrupted resources would be more as compared to the number of resources affected in a less significant disruption. Since most of the short haul flights operate during daytime, disruptions which occur early in a day, tend to have more effect in the schedule as opposed to those which occur later in the day. If the disrupted flight is the last flight of the day, then neighborhood generation is not required. In this case, not only would the delay wash-out automatically in the schedule, but also since there is no flight to operate afterwards the need for schedule recovery does not arise.

Disruptions which occur at ports with less frequent in-bound and out-bound flights take longer to resolve. During schedule recovery, the time window (length of the recovery period) defines the inclusion of disrupted resources in the neighborhood, and it needs to be sufficiently long to achieve a feasible solution by end of the day's operations. At the same time, the recovery window is desired to be as short as possible to get back to the original schedule in the minimum possible time. This approach is referred to as a *disruption neighborhood*. Therefore, factors that are critical for the generation of the disruption neighborhood are as follows:

- Time of disruption.
- Port of disruption.
- Ground time available between two consecutive tasks for a resource included in the disruption neighborhood.
- Resource availability (maintenance/reserve/overtime).

In the neighborhood, each resource is given entry and exit conditions. The entry and exit conditions include an 'entry time', 'exit time', 'entry port', and 'exit port' within which the resource can perform any legal duty or task that is given to them in the recovery period. Entry condition is determined by the 'time' and 'port' a resource is available to perform a task at in the neighborhood. Available time is the time when a resource can perform the next task (arrival time + minimum turnaround time or minimum sit time). Whereas, the exit condition is determined by the first task a resource is assigned to perform outside the recovery period. Flights that are affected by the disruption and are no longer feasible to operate as scheduled are included in the disruption neighborhood. In other words, those flights are included in the neighborhood whose corresponding disrupted resources are unable to operate their scheduled duties. The process of identification of disrupted resources and tasks is explained in detail in the following sections.

8.3 INITIAL NEIGHBORHOOD

Disruption neighborhood is generated in steps. First, an initial neighborhood is generated in which the idea is to recover the schedule with as little changes to the planned schedule as possible. If it is not possible to get a feasible recovery solution, then the neighborhood is extended by including more resources (and tasks) to get back to scheduled operations. This process continues till operations are back to the originally planned schedule or the end of the day's operations are reached. The disruption neighborhood is generated in three steps.

1. Recovery period selection.
2. Inclusion of resources in the neighborhood.
3. Tasks included in the neighborhood.

Step-1: Initial Recovery Period (Time Window) Selection

Initial recovery period is the time difference of disruption start time and disruption end time. Disruption start time is the first instance when the disruption is known. Whereas, disruption end time is determined either by the start time of the uncovered task or by the time of availability of a needed resource. In other words, the recovery period for the initial neighborhood is selected from the time disruption is known until the departure time of the uncovered flight (which cannot depart as scheduled due to unavailable required resource). A required resource is the one that is needed to cover the uncovered flight. If the resources of the disrupted flight are scheduled to do different tasks, then the recovery period will be extended until the start time of the first task which happens later in time in the schedule. This determines the initial recovery period of the neighborhood.

Step-2: Inclusion of Resources in the Neighborhood

a. Resources of the disrupted flight, if they have scheduled task(s) in the network.
b. Resource(s) which are idle at the port of disruption and which can be utilized to cover the uncovered task.
c. Resources of the flight(s) included in the neighborhood.

Step-3: Tasks Included in the Neighborhood

a. The uncovered flight.
b. The flight(s), that initiate from the port of disruption and falls in the initial tie window.
c. Flights(s) corresponding to the resource(s) in Step-2c.

When a resource is included in the neighborhood, they are considered as a potential candidate subject to the degree of unattractiveness of their recovery duties to cover the uncovered tasks included in the neighborhood. To describe the application of the generation of disruption neighborhood approach, a schedule recovery problem instance for a single fleet type is presented and explained in detail. In the problem instance as shown in Figure 8.1, flight 527 departs from Delhi (DEL) at 1300 h and is scheduled to arrive in Ahmedabad (AMD) at 1420 h. However, due to the delay in the landing clearance by the ATC at Ahmedabad, the flight is delayed by 30 min and arrives in AMD at 1450 h. The disruption was known at 1400 h while flight 527 was airborne and approaching its destination. The aircraft (A1) of flight 527 is next scheduled to do flight 520 at 1510 h from AMD but due to the delay it cannot operate flight 520 as scheduled because the minimum turn around condition (minimum 30 min are required between two consecutive duties for an aircraft) will be violated. In this scenario, to cover the disrupted flight 520, the needed resource is an aircraft. For this, the disrupted schedule is required to be back to normal operations by generating the initial neighborhood at the port of disruption (AMD).

In the example, flight 520 is uncovered due to the delayed arrival of the aircraft of flight 527 at AMD. Hence, the initial disruption neighborhood is required to be generated at AMD within the initial recovery period. Therefore, the first step is to identify the initial recovery period by identifying the time window within which the schedule is expected to be recovered. While generating the initial neighborhood, the cause of disruption in the schedule is critical in analyzing and anticipating the impact of disruption. In Figure 8.1, the initial recovery period is 1 h and 10 min and the cause of disruption is the unavailability of an aircraft.

The second step in generating the initial neighborhood is to look for the resource(s) in the initial recovery period at the port of disruption (AMD) which may help the uncovered flight depart. At this step, inclusion of idle resources, if any, is preferred as opposed to the resources which are already assigned a flight task in the original schedule. Also, resources (aircraft in this case) at the port of disruption (AMD) which

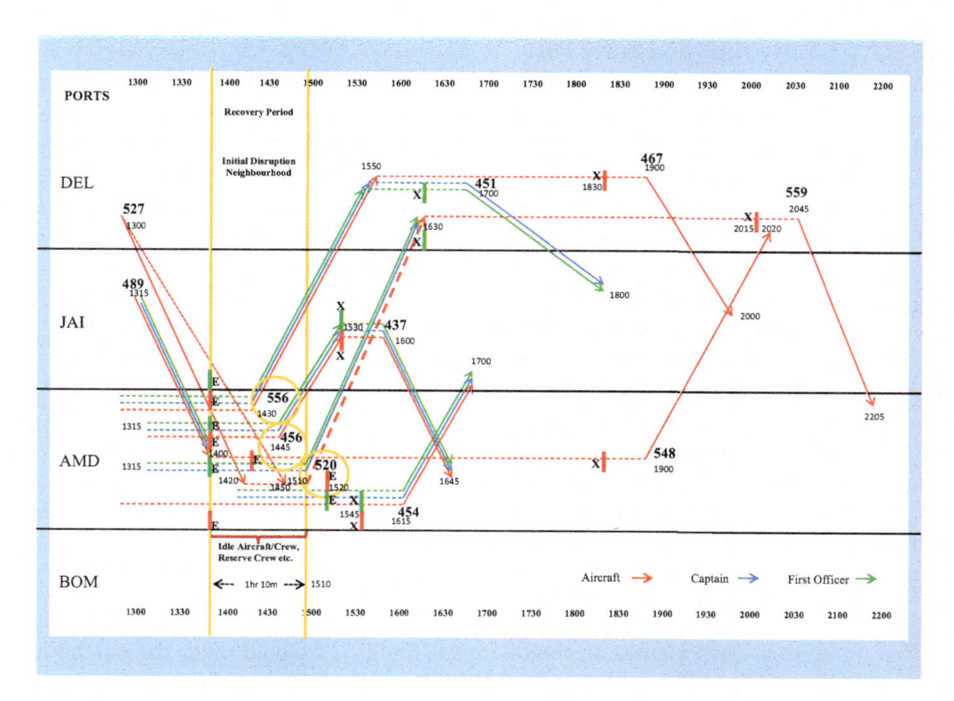

FIGURE 8.1 Initial disruption neighborhood.

are scheduled to operate other flights and are departing from the port of disruption within the initial recovery period are included in the neighborhood. Therefore, based on the scenario presented in Figure 8.1, for the generation of the initial neighborhood, two idle aircraft and two aircraft which are scheduled to depart on flights 556 (aircraft A2) and 456 (aircraft A3) from AMD are included within the initial recovery period.

In the third step, flights 556 and 456 along with the crew, i.e., captain (C) and the first officer (F) of these flights are also included in the initial neighborhood because without aircraft these flights will become uncovered and hence cannot operate as planned and become a part of the initial disruption neighborhood.

To generate the initial neighborhood, the entry and exit conditions are given to all the resources. In the ground disruptions, neighborhood is generated at the departure port of the disrupted flight. For this type of disruption, it is desirable to know the cause of disruption, as 'entry times' for the resources into the neighborhood may be different from one resource to another. Whereas, for the disruptions that are acknowledged mid-air, the precise time of disruption is important to know to get the 'entry times'. In this scenario, most of the resources included in the neighborhood would get identical initial 'entry times'.

8.3.1 Resources of the Disrupted Flight: Aircraft

Aircraft (A1) was scheduled to operate flight 527 and a delayed arrival of this flight would prohibit A1 to do its next flight 520 as planned in the original schedule.

8.3.1.1 Entry Time and Entry Port

Aircraft A1 was scheduled to arrive at AMD at 1420 h but with a delay of 30 min its actual arrival time was 1450 h. As per the rules, it is mandatory that between two consecutive flights a minimum of 30-min turnaround (TA) duration is given to a resource before it can perform the next task/duty. Therefore, after the actual arrival time (1450 h) of the aircraft (A1) and with a mandatory ground time, A1 would be available to do the next flight at 1520 h. As a result, A1 would enter the disruption neighborhood at 1520 h at AMD.

8.3.1.2 Exit Time and Exit Port

After the disrupted flight 527, A1 was scheduled to operate flight 520. However, due to delay it would not be able to operate flight 520 on time but it can still operate flight 559 from DEL on time (2045 h) since by then, delay is washed out in the schedule. So, considering the minimum turnover condition for the aircraft, the exit time and exit port for A1 would be at 2015 h in DEL.

8.3.2 Resources of the Disrupted Flight: Crew

8.3.2.1 Entry Time and Entry Port

For any flight, the aircraft and crew have identical departure and arrival times. Therefore, crew associated with the disrupted flight 527 would enter the neighborhood at the same time as the disrupted aircraft (A1). Hence, captain (C1) and first officer (F1) would also enter the neighborhood at 1520 h at AMD. Since the flight 527 is not the last flight of the day, therefore, the flights subsequently affected by the disruption are also considered. The next flight for aircraft A1 is to operate flight 520 however, the next flight assigned for the crew of the disrupted flight 527 is flight 454. Unlike *unit-crewing*, this is a situation where resources split and do not remain together to operate their subsequent flight in the schedule. Whereas, in unit-crewing, crew and aircraft stay together to perform flight tasks.

8.3.2.2 Exit Time and Exit Port

The captain C1 and the first officer F1 of the disrupted flight 527 are scheduled to operate flight 454 from AMD. The delay was 30 minutes due to which crew of the disrupted flight 527 arrive at the port (AMD) at 1450 h. After the 30-min sit time condition is satisfied, the crew would be available to do their next flight in the sequence at 1520 h. However, the next flight 454 in the schedule for crew (C1 and F1) is at 1615 h. The ground time for crew in the schedule is sufficient to wash-out the delay of 30 min hence the crew can perform their duties as planned in the original schedule. Hence, exit time and exit port for C1 and F1 would be 1545 h at AMD.

8.3.3 Resources of the Affected Flights

Based on the recovery time window of 1 h 10 min, all the flights departing from AMD are included in the neighborhood. The flights departing from AMD during the recovery time window are flights 520 (AMD-DEL), 456 (AMD-JAI), and 556 (AMD-DEL). As a result, the resources, i.e., aircraft, captain, and first officer of these flights would also enter the disruption neighborhood. In Table 8.1, all the resources and

TABLE 8.1
Resources and flights in the initial neighborhood

Neighborhood Resources	Reason of Inclusion	Resource				Neighborhood Flights					Reason of Inclusion
		Entry		Exit			Departure		Arrival		
		Port	Time	Port	Time	No.	Port	Time	Port	Time	
A1 C1 F1	On the disrupted flight	AMD	1400	DEL	2015						
		AMD	1400	AMD	1545						
		AMD	1400	AMD	1545						
A1 (already in) C2 F2	Inclusion of flight 520	AMD	1400	DEL	2015	520	AMD	1510	DEL	1630	Aircraft unavailability
		AMD	1400	DEL	1630						
		AMD	1400	DEL	1630						
A2 C3 F3	Inclusion of flight 556	AMD	1400	DEL	1830	556	AMD	1430	DEL	1550	Departing inside time window
		AMD	1400	DEL	1630						
		AMD	1400	DEL	1630						
A3 C4 F4	Inclusion of flight 456	AMD	1400	JAI	1530	456	AMD	1445	JAI	1530	Departing inside time window
		AMD	1400	JAI	1530						
		AMD	1400	JAI	1530						
A4	Idle inside disruption window	AMD	1430	AMD	1830						
A5	Idle inside disruption window	AMD	1400	AMD	1545						

flights included in the initial neighborhood are given their entry and exit conditions along with their reason of inclusion in the neighborhood.

After generating the initial neighborhood, the optimization problem is solved with the aim to minimize the deviation (i.e., cost) from the originally planned schedule. If a feasible solution is achieved, the schedule is recovered. Otherwise, the initial neighborhood is expanded to include more resources in the neighborhood as explained in the following section.

8.4 EXPANSION OF NEIGHBORHOOD

Expansion of the neighborhood is required if the recovery problem is infeasible in the initial neighborhood, i.e., if there are insufficient resources which can cover all the tasks in it, then the initial neighborhood is modified by expanding the initial recovery period. This can be achieved in the following way. After the departure time of the uncovered flight, i.e., end of the initial recovery period, the first opportunity to get the needed resource (aircraft in this case) is looked at the port of disruption. Due to this resource, the flight is uncovered, and recovery period is extended until the time of availability of the desired resource. During this process, the needed resource and other resources (captain and first officer in this case) involved with the task(s) are included in the neighborhood.

In the scenario presented in Figure 8.1, flight 520 is delayed by the time the next aircraft is available at AMD. An illustration of expanded neighborhood is presented

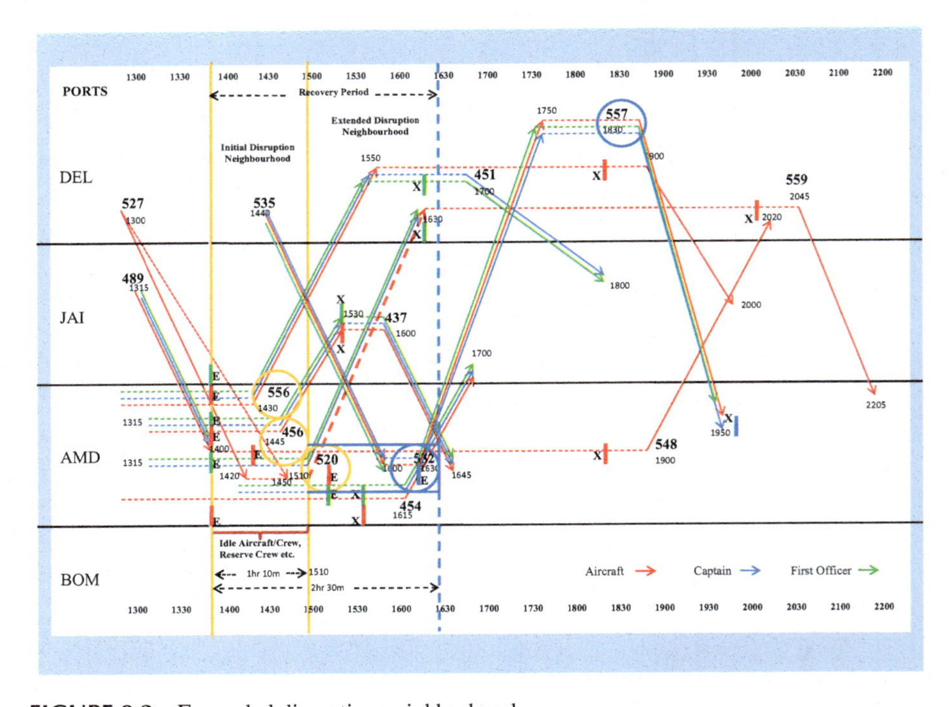

FIGURE 8.2 Expanded disruption neighborhood.

TABLE 8.2
Disrupted resources and the flights in the neighborhood

		Initial Neighborhood Flights and Resources									
		Resource				Neighborhood Flights					
		Entry		Exit			Departure		Arrival		
Neighborhood Resources	Reason of Inclusion	Port	Time	Port	Time	No.	Port	Time	Port	Time	Reason of Inclusion
A1 C1 F1	On the disrupted flight	AMD	1520	DEL	2015						
		AMD	1520	AMD	1545						
		AMD	1520	AMD	1545						
A1 (already in) C2 F2	Inclusion of flight 520	AMD	1400	DEL	2015	520	AMD	15:10	DEL	16:30	Aircraft unavailability
		AMD	1400	DEL	1630						
		AMD	1400	DEL	1630						
A2 C3 F3	Inclusion of flight 556	AMD	1400	DEL	1830	556	AMD	14:30	DEL	15:50	Departing inside window
		AMD	1400	DEL	1630						
		AMD	1400	DEL	1630						
A3 C4 F4	Inclusion of flight 456	AMD	1400	JAI	1530	456	AMD	14:45	JAI	15:30	Departing inside window
		AMD	1400	JAI	1530						
		AMD	1400	JAI	1530						
A4	Idle inside disruption window	AMD	1430	AMD	1830						
A5	Idle inside disruption window	AMD	1400	AMD	1545						

(continued)

TABLE 8.2 (Continued)
Disrupted resources and the flights in the neighborhood

		Initial Neighborhood Flights and Resources									
		Resource				Neighborhood Flights					
		Entry		Exit			Departure		Arrival		
Neighborhood Resources	Reason of Inclusion	Port	Time	Port	Time	No.	Port	Time	Port	Time	Reason of Inclusion
		Expanded Neighborhood Flights and Resources									
A6 C5 F5	Inclusion of flight 532	AMD	1630	DEL	1800	532	AMD	16:30	DEL	17:50	Aircraft borrowed to cover flight
		AMD	1630	DEL	1800						
		AMD	1630	DEL	1800						
A3 (already in) C4 (already in) F4 (already in)		AMD	1400	AMD	1715	437	JAI	16:00	AMD	16:45	Aircraft borrowed to cover flight
		AMD	1400	AMD	1715						
		AMD	1400	AMD	1715						
A5 (already in) C6 F6	Reserve Reserve	AMD	1400	JAI	1730	454	AMD	16:15	JAI	17:00	Aircraft borrowed to cover flight
		AMD	1400	AMD	2000						
		AMD	1400	AMD	2000						

in Figure 8.2 in which flight 535 arrives in AMD at 1600 h. The aircraft of this flight is scheduled to do flight 532 as its next task at 1630 h. The neighborhood is expanded until 1630 h to include the aircraft of flight 535, and therefore, flight 532 is also included in the neighborhood along with its crew since the aircraft of flight 532 can be used for another flight and flight 532 cannot operate without aircraft and would become uncovered. As a result, entry and exit conditions are given for the aircraft and the crew of flight 532. Based on this, as shown in Figure 8.2, the recovery window is extended to 2 h and 30 min and the downstream flights for the involved resources in the expanded neighborhood are included. This process continues until the schedule is back to the planned operations.

In Table 8.2, all disrupted resources and tasks included in the expanded neighborhood are presented along with the reason for their inclusion in the neighborhood.

8.5 NETWORK CREATION AND COLUMN GENERATION

To solve the schedule disruption problem, a network of the resources (with their entry and exit conditions) and the flights included in the neighborhood is created. In the network, a node represents entry and exit conditions for a resource and a flight. Arcs are created between; entry nodes and flight nodes, two flight nodes (if need be), and flight nodes and exit nodes. The process of generating the disruption network by creating the arcs is presented in Figure 8.3. After generating all the possible arcs in the network, paths are generated. A path is a sequential set of arcs connecting entry and exit nodes of a given resource to cover the flights included in the neighborhood.

A path is created if the following explicit and implicit conditions are satisfied.

- Port conditions (*explicit condition*)
  The entry port for a resource must be identical to the departure port of the flight. Similarly, the exit port of a resource must also be the same as the arrival port of the flight.
- Time condition (*explicit condition*)
  The entry time of a resource must be less than or equal to the flight departure time. Also, the exit time of the resource must be greater or equal to the arrival time of the flight.
- Identity condition (*implicit condition*)
  The identity of a resource must remain the same throughout the path, i.e., at the entry node and the exit node.

Through this, columns are created. Mathematically, a path is known as a column. Column generation is a technique in which columns are generated to solve optimization problems using sophisticated optimization solvers such as CPLEX or Gurobi.

In column generation, a column corresponds to a sequence of tasks a resource can do. In large optimization problems, there can be numerous columns in a problem. For large linear programming problems, not all columns are considered at once for two major reasons. First, most of the variables in the problem would be non-basic, i.e., they are not in the basis. Even though the number of variables may be large, only a

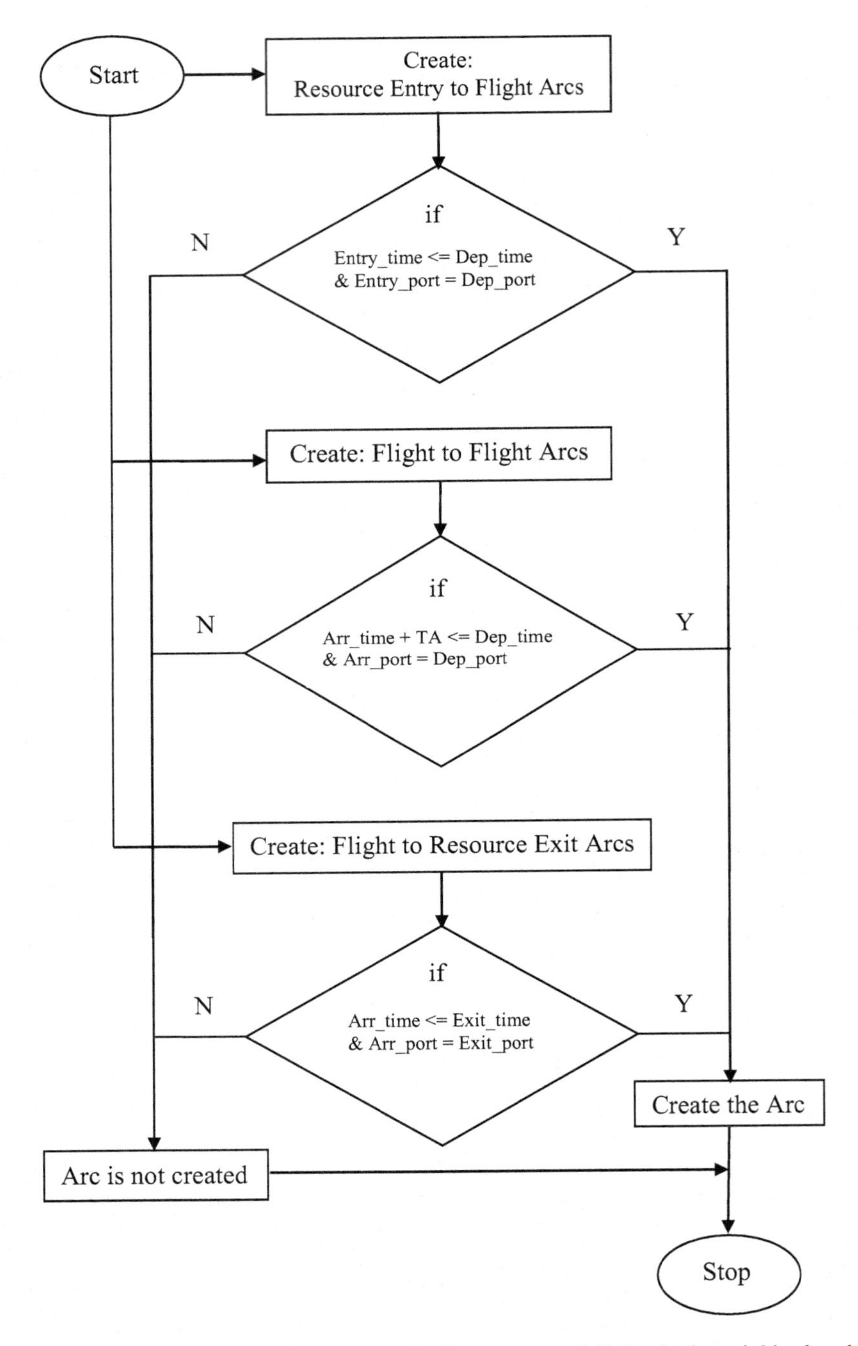

FIGURE 8.3 Flow chart to create a network of resources and flights in the neighborhood.

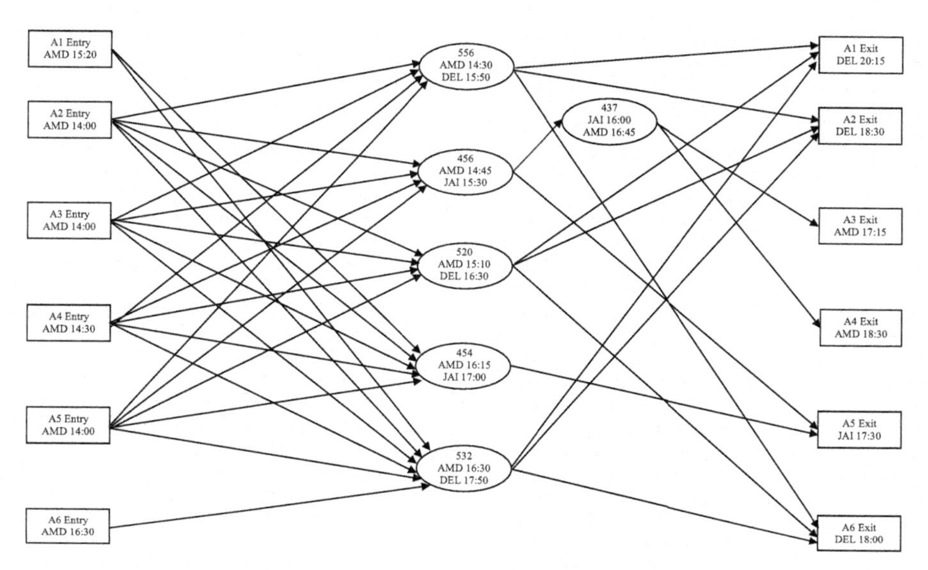

FIGURE 8.4 Arcs connecting aircraft with uncovered flights in the neighborhood.

small subset of the decision variables appears in the optimal solution. Second, since the number of variables could be extremely large, therefore, computationally it would be unnecessary and time consuming to consider all the variables without improving the objective function value. In such cases the problem is divided into the master problem and the sub-problem. The master problem is the original problem in which initially only a subset of variables is considered. However, the sub-problem identifies the new variable with negative reduced cost to improve the objective function value. For problems with a smaller number of variables such as schedule recovery problems all the columns can be generated together without compromising either on computational time or quality of the solution. In the network created in Figure 8.4, the problem size is small since the number of tasks and resources in the disruption neighborhood are few. As a result, number of columns generated is significantly less than the number of columns generated for schedule planning problems such as crew scheduling where number of possible columns could be significantly large.

8.5.1 Aircraft Network

As presented in Figure 8.3, based on the entry and exit conditions for each aircraft and flight included in the neighborhood, aircraft arcs are generated as shown in Figure 8.4.

8.5.1.1 Aircraft Routing

A broken aircraft routing is presented in Figure 8.5 representing entry and exit conditions for aircraft A1 and the disrupted flight 520.

For aircraft A1, a path cannot be created due to the violation of *time condition* since the entry time (1520 h) of A1 is after the departure (1510 h) of flight 520.

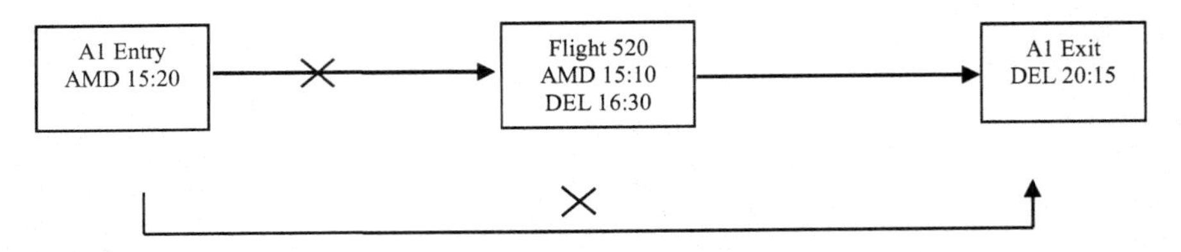

FIGURE 8.5 A broken aircraft routing.

TABLE 8.3
Task and constraints for aircraft A1

		X_1		
Aircraft	A1	1	=	1
Tasks	A1 Entry	1	=	1
	A1 Exit	1	=	1
	Flight 520	0	=	1

However, there is an arc between flight node of flight 520 and exit node of A1 but aircraft A1 cannot operate flight 520, therefore this arc bears no meaning, and the path is incomplete. Table 8.3 represents aircraft and flight constraints along with the column representation of the broken routing or infeasible path for aircraft A1. A path or column in the network is represented by a variable (x_i) in the problem formulation. The right-hand side value ensures that all the tasks within the recovery period are covered exactly once.

In this scenario, either flight 520 needs to be delayed by at least 10 min for aircraft A1 to operate it at 1520 h or another aircraft needs to be made available to cover for it so that flight 520 leaves at 1510 h.

8.5.1.2 Aircraft Columns

Based on the network presented in Figure 8.3, columns are generated for all aircraft included in the neighborhood (Table 8.4). Each column is represented by a variable (xi).

In Table 8.4, the rows represent aircraft and flight constraints. The columns indicate that aircraft A1 can only operate flight 532. However, for aircraft A2, there are three options, i.e., flight 556, flight 520, and flight 532 to cover in the neighborhood. There are two options to operate flight 456 and flight 437, i.e., through aircraft A3 and aircraft A4, respectively. Aircraft 5 can cover two flights, i.e., flight 456 and flight 454 but aircraft A6 can only operate flight 532. As evident from Table 8.4, each flight is covered at least once by the aircraft, i.e., the aircraft are available to cover all six flights which are included in the neighborhood. Based on the cost associated with each column, the least expensive column will be selected by the optimizer to minimize the total cost of the schedule recovery problem.

TABLE 8.4
Aircraft columns

		X_1	X_2	X_3	X_4	X_5	X_6	X_7	X_8	X_9		
Aircraft	A1 Entry	1	0	0	0	0	0	0	0	0	=	1
	A1 Exit	1	0	0	0	0	0	0	0	0	=	1
	A2 Entry	0	1	1	1	0	0	0	0	0	=	1
	A2 Exit	0	1	1	1	0	0	0	0	0	=	1
	A3 Entry	0	0	0	0	1	0	0	0	0	=	1
	A3 Exit	0	0	0	0	1	0	0	0	0	=	1
	A4 Entry	0	0	0	0	0	1	0	0	0	=	1
	A4 Exit	0	0	0	0	0	1	0	0	0	=	1
	A5 Entry	0	0	0	0	0	0	1	1	0	=	1
	A5 Exit	0	0	0	0	0	0	1	1	0	=	1
	A6 Entry	0	0	0	0	0	0	0	0	1	=	1
	A6 Exit	0	0	0	0	0	0	0	0	1	=	1
Task	Flight 556	0	1	0	0	0	0	0	0	0	=	1
	Flight 456	0	0	0	0	1	1	1	0	0	=	1
	Flight 437	0	0	0	0	1	1	0	0	0	=	1
	Flight 520	0	0	1	0	0	0	0	0	0	=	1
	Flight 454	0	0	0	0	0	0	0	1	0	=	1
	Flight 532	1	0	0	1	0	0	0	0	1	=	1

8.5.2 Captain Network

As presented in Figure 8.3, based on the entry and exit conditions for each captain and flight included in the neighborhood, arcs are generated as shown in Figure 8.6. For instance, based on the neighborhood entry condition for captain C1, the captain can potentially be assigned flight 454 or flight 532. However, there is no flight that can get C1 back to AMD on or before 1545 h and satisfy the exit condition for captain C1. This indicates that C1 will not be assigned any flights in the neighborhood. In Figure 8.6, this scenario is represented by a dashed arrow between the entry and exit conditions for captain C1. However, for other captains included in the disruption neighborhood, their entry and exit conditions are satisfied while covering the disrupted flights. As evident from Figure 8.6, flight 437 is not covered by any captain. In the network, there is no direct arc between the entry node of a captain and the node for flight 437. Also, there is no captain who can do flights 456 and 437 in sequence after satisfying their entry and exit conditions.

8.5.2.1 Captain Pairings

Crew pairings for captains C2 and C3 are presented in Figure 8.7 representing their entry and exit conditions and the disrupted flight 520.

For C2 there is a path from its entry node to its exit node through flight node 520. Similarly, for C1 there exists a path from its entry node to exit node. However, there

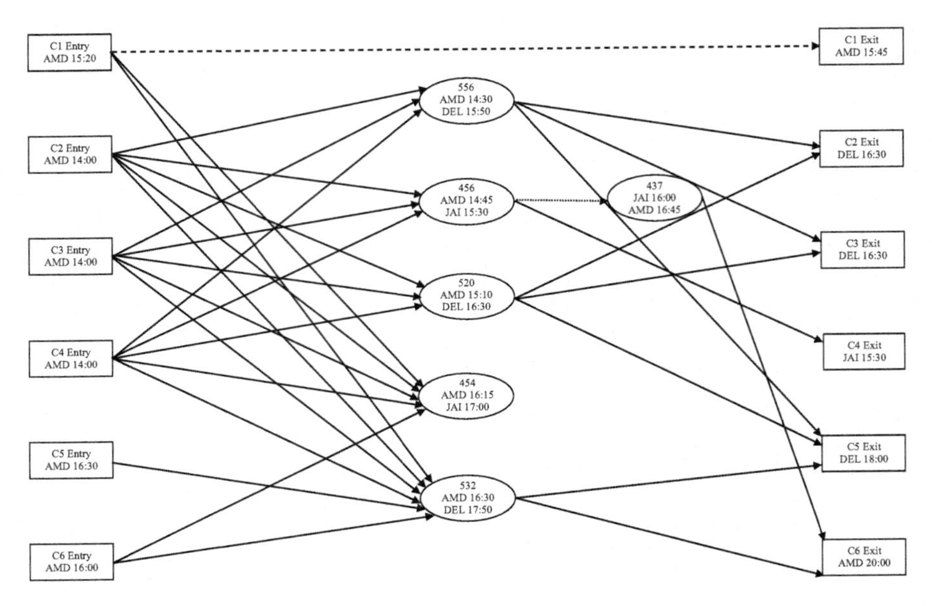

FIGURE 8.6 Arcs connecting captain with uncovered flights in the neighborhood.

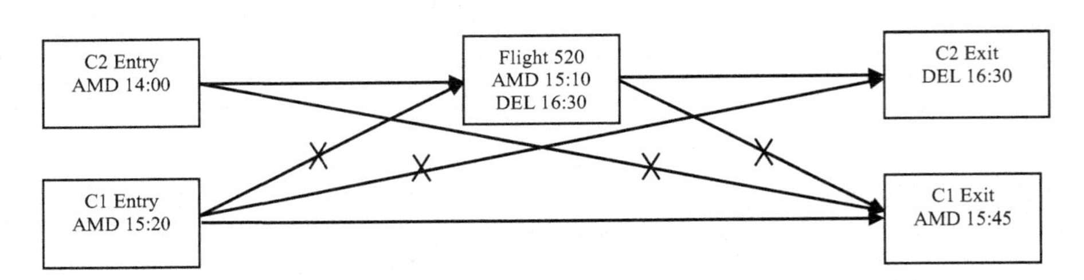

FIGURE 8.7 Broken pairing for the captain.

is no path from the entry node of C2 to the exit node of C1 as it violates *identity condition.* There exists no path from entry node of C1 to exit node of C1 through flight node 520 as it violates *time and port conditions.*

8.5.2.2 Captain Columns

Table 8.5 represents resource constraints and task constraints along with column representation for the feasible paths for captains C1 and C2. For captain, C2, a column is generated since the captain can operate flight 520 as per the scheduled departure time (1510 h) of the flight. For captain C1, a column cannot be generated as they cannot operate flight 520 at the scheduled departure time. Captain C1 can cover flight 520 if it is delayed by at least 10 min, in which case a column for C1 can also be generated. All captain columns are presented in Table 8.6.

TABLE 8.5
Task and resource constraints for captions

		X_1	X_2		
Captains	C2	1	0	=	1
	C1	0	1	=	1
Tasks	C2 Entry	1	0	=	1
	C1 Entry	0	1	=	1
	C2 Exit	1	0	=	1
	C1 Exit	0	1	=	1
	Flight 520	1	0	=	1

TABLE 8.6
Captain columns

		X_1	X_2	X_3	X_4	X_5	X_6	X_7	X_8	X_9	X_{10}	X_{11}	X_{12}	X_{13}		
Captain	C1 Entry	0	0	0	0	0	0	0	1	0	0	0	0	0	=	1
	C1 Exit	0	0	0	0	0	0	0	1	0	0	0	0	0	=	1
	C2 Entry	1	1	0	0	0	0	0	0	1	0	0	0	0	=	1
	C2 Exit	1	1	0	0	0	0	0	0	1	0	0	0	0	=	1
	C3 Entry	0	0	1	1	0	0	0	0	0	1	0	0	0	=	1
	C3 Exit	0	0	1	1	0	0	0	0	0	1	0	0	0	=	1
	C4 Entry	0	0	0	0	1	0	0	0	0	0	1	0	0	=	1
	C4 Exit	0	0	0	0	1	0	0	0	0	0	1	0	0	=	1
	C5 Entry	0	0	0	0	0	1	0	0	0	0	0	1	0	=	1
	C5 Exit	0	0	0	0	0	1	0	0	0	0	0	1	0	=	1
	C6 Entry	0	0	0	0	0	0	1	0	0	0	0	0	1	=	1
	C6 Exit	0	0	0	0	0	0	1	0	0	0	0	0	1	=	1
Task	Flight 556	1	0	1	0	0	0	0	0	0	0	0	0	0	=	1
	Flight 456	0	0	0	0	1	0	0	0	0	0	0	0	0	=	1
	Flight 437	0	0	0	0	0	0	0	0	0	0	0	0	0	=	1
	Flight 520	0	1	0	1	0	0	0	0	0	0	0	0	0	=	1
	Flight 454	0	0	0	0	0	0	0	0	0	0	0	0	0	=	1
	Flight 532	0	0	0	0	0	1	1	0	0	0	0	0	0	=	1

8.5.3 First Officer Network

As presented in Figure 8.3, based on the entry and exit conditions for each first officer and flight included in the neighborhood, possible arcs are generated as shown in Figure 8.8.

For instance, based on the neighborhood entry condition for first officer F1, the first officer can either be assigned flight 454 or flight 532. However, there is no flight that can satisfy the exit condition for first officer F1 in the limited window of 25 min

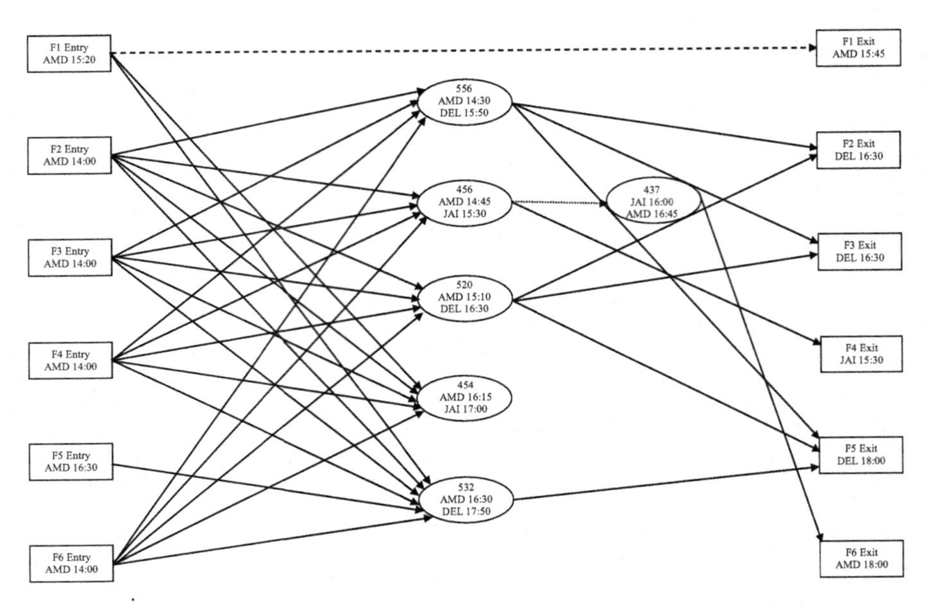

FIGURE 8.8 First officer arcs in the recovery neighborhood.

between the entry and exit times for F1. This indicates that first officer F1 will not be assigned any flights in the neighborhood unless F1 has sufficient time duration between the entry and exit time to cover any flight in the network.

In Figure 8.8, a dashed arrow between the entry and exit conditions for first officer F1 represents that this resource is on the ground and cannot be used for recovery operations. Contrarily, the entry and exit conditions of other first officers included in the disruption neighborhood are satisfied while covering the disrupted flights. For instance, first officers F2, F3, and F4 can potentially be assigned flights 556, 456, 520, 454, and 532. Based on the entry conditions (time and port), first officers F2, F3, and F4 satisfy the flight departure schedule. However, to get back to their required exit ports, first officers F2, F3, and F4 can only be assigned the flights which would get them back to their required port on or before the required time to satisfy their respective exit conditions. For this reason, fewer arcs emanate from the flight nodes to the exit nodes as opposed to the number of arcs emanating from the entry nodes to the flight nodes. Furthermore, as evident from Figure 8.8, flight 437 is not covered by any first officer. In the network, there is no direct arc between the entry node of a first officer and the node for flight 437. Also, there is no first officer who can do flights 456 and 437 in sequence after satisfying their entry and exit conditions.

8.5.3.1 First Officer Pairings

Crew pairings for first officers F2 and F3 are presented in Figure 8.9 representing their entry and exit conditions and the disrupted flight 520.

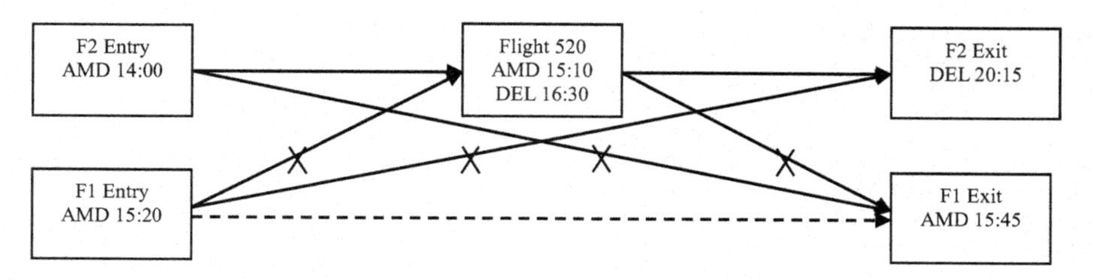

FIGURE 8.9 Broken pairing for the first officer.

TABLE 8.7
Task and resource constraints for first officers

		X_1	X_2		
First Officers	F2	1	0	=	1
	F1	0	1	=	1
Tasks	F2 Entry	1	0	=	1
	F1 Entry	0	1	=	1
	Flight 520	1	0	=	1
	F2 Exit	1	0	=	1
	F1 Exit	0	1	=	1

For first officer F2, there is a path from its entry node to its exit node. This means that, based on the entry and exit conditions, the first officer F2 can cover flight 520 and get back to the required port (DEL) by 2015 h. Similarly, for F1 there exists a path from its entry node to its exit node. However, this path indicates that the resource is on the ground and cannot be used for recovery operations. This is represented through a dashed arrow between the entry and exit conditions for first officer F1. However, there is no path from the entry node of F2 to the exit node of F1 as it violates the *identity condition.* It is important to ensure that resource identity conditions are imposed throughout the network creation. This aspect is critical in creating feasible recovery duties in the disruption neighborhood. Also, there exists no path from the entry node of F1 to its exit through flight 520 as it violates *time conditions*. However, first officer F1 can cover flight 520 if it is delayed by at least 10 min, in which case a column for F1 can also be generated.

Table 8.7 represents first officer constraints and task constraints along with the column representation for the feasible paths for first officers F2 and F1. For first officer, F2, a column is generated since the officer can operate flight 520 as per the scheduled departure time (1510 h) of the flight. For first officer F1, a column cannot be generated as they cannot operate flight 520 at the scheduled departure time. First officer F1 can cover flight 520 if it is delayed by at least 10 min, in which case a column for F1 can also be generated. At this stage, apart from other reasons, the problem is infeasible since aircraft (A1) available in the neighborhood cannot operate flight 520 as per the schedule. To get to a feasible solution, either the neighborhood needs to be expanded or flight 520 is to be delayed by 10 min so that A1 can operate flight 520.

8.5.3.2 First Officer Columns

In Table 8.8, columns are generated for all first officers included in the disruption neighborhood.

Columns x1 to x7 indicate the paths created for the first officers covering at least one flight in the neighborhood. Whereas, columns x8 to x13 represents the null variables which means that there is only entry and exit path for the resource and no flight is covered in the path. This scenario is expensive for the airline, indicating that the resource is idle and not being utilized.

In airline schedule recovery scenario, not all planning constraints which are relevant in the schedule planning stage for resources are taken into consideration. Aircraft constraints such as *through connection* (two flights must be operated in sequence by the same aircraft) or *flying time,* i.e., all aircraft routings to contain roughly the same number of flights and the same amount of flying time are not relevant from the recovery perspective. Similarly, fuel costs have already been calculated in the construction of the schedule, therefore during recovery it is not significant which aircraft is assigned which routings in the single fleet recovery.

Nevertheless, planning constraints such as *maintenance limit* (each aircraft must have a maintenance check every 36 h) and *minimum-turn-time* (minimum 30 min are required between two consecutive duties for each aircraft) are ensured even during the generation of recovery schedule. For crew, the implicit constraints such as legalities surrounding the crew duties are incorporated during column generation while generating the feasible recovery duties for the resources included in the neighborhood

TABLE 8.8
First officer columns

		X_1	X_2	X_3	X_4	X_5	X_6	X_7	X_8	X_9	X_{10}	X_{11}	X_{12}	X_{13}		
First Officer	F1 Entry	0	0	0	0	0	0	0	1	0	0	0	0	0	=	1
	F1 Exit	0	0	0	0	0	0	0	1	0	0	0	0	0	=	1
	F2 Entry	1	1	0	0	0	0	0	0	1	0	0	0	0	=	1
	F2 Exit	1	1	0	0	0	0	0	0	1	0	0	0	0	=	1
	F3 Entry	0	0	1	1	0	0	0	0	0	1	0	0	0	=	1
	F3 Exit	0	0	1	1	0	0	0	0	0	1	0	0	0	=	1
	F4 Entry	0	0	0	0	1	0	0	0	0	0	1	0	0	=	1
	F4 Exit	0	0	0	0	1	0	0	0	0	0	1	0	0	=	1
	F5 Entry	0	0	0	0	0	1	0	0	0	0	0	1	0	=	1
	F5 Exit	0	0	0	0	0	1	0	0	0	0	0	1	0	=	1
	F6 Entry	0	0	0	0	0	0	1	0	0	0	0	0	1	=	1
	F6 Exit	0	0	0	0	0	0	1	0	0	0	0	0	1	=	1
Task	Flight 556	1	0	1	0	0	0	0	0	0	0	0	0	0	=	1
	Flight 456	0	0	0	0	1	0	1	0	0	0	0	0	0	=	1
	Flight 437	0	0	0	0	0	0	1	0	0	0	0	0	0	=	1
	Flight 520	0	1	1	0	0	0	0	0	0	0	0	0	0	=	1
	Flight 454	0	0	0	0	0	0	0	0	0	0	0	0	0	=	1
	Flight 532	0	0	0	0	0	1	0	0	0	0	0	0	0	=	1

and are not stated explicitly. For instance, constraints such as *minimum-sit-time* (minimum 30 min required between two consecutive flights) and *base constraints* (all crew pairings must start and end at the same base) are valid during schedule recovery operations. However, for captains and first officers there are few planning constraints which are not considered during recovery process such as *maximum sectors* (each duty period to contain a maximum number of flights). A crew can perform any task(s) as long as their recovery duties are feasible in the neighborhood.

8.6 SET PARTITIONING FORMULATION

Mathematically, the airline schedule recovery problem can be formulated as a set partitioning problem as presented in Table 8.9.

The objective of the recovery problem is to find a minimum cost set of feasible recovery duties for all the resources in the disruption neighborhood such that all the tasks within the recovery period are covered. In Table 8.9, the resource constraints ensure that each resource is assigned to exactly one recovery duty. Resource constraints have a generalized upper bound structure because the constraints for each resource are disjointed, and each column represents exactly one task for a resource. The task constraints ensures that each task (flight in this case) in the recovery schedule is covered exactly once in the recovery neighborhood by an aircraft, a captain, and a first officer. To achieve this, for each flying task, a copy corresponding to aircraft, captain, and first officer is generated. Similarly, task constraints ensure that each task in the recovery network is covered exactly once by an aircraft, captain, and first officer.

Null variable(s) in the solution indicate that the resource is idle and does not perform any flying duty in the neighborhood. Depending on the location of the resource and the port of disruption, idle resource(s) can be assigned a non-flying duty in the recovery neighborhood. Similarly, the presence of artificial variables in the solution indicates the infeasibility of the solution. In case of an infeasible solution, the neighborhood needs to be expanded by including more resources in the existing set of resources to get the schedule back to the planned operations. All tasks in the recovery network, are assigned a corresponding artificial variable to identify the reason of infeasibility of the solution so that the appropriate resource can be included in the disruption neighborhood. The process of neighborhood expansion continues until artificial variables do not appear in the solution. In schedule recovery, a solution that contains either decision variables or null variables is considered feasible. The null variables and the artificial variables take the form of identity matrix (I).

8.7 ROLLING TIME HORIZON RECOVERY

After a disruption problem is solved and operations are back as scheduled then more disruptions may occur leading to the breakdown of the schedule again. Also, it is possible that while schedule is being recovered another disruption happens in the same time window. In such situations, the schedule recovery problem is solved more than once, and the solution is updated with each problem instance. This process is called *rolling time horizon recovery*. The process of rolling time horizon is illustrated below.

TABLE 8.9
A set partitioning representation depicting resource and task constraints for a schedule recovery problem

	x_1^1	x_2^1	x_3^1	x_4^1	x_5^2	x_6^2	x_7^3	x_8^3	x_9^1	x_{10}^1	x_{11}^1	x_{12}^2	x_{13}^2	x_{14}^2	x_{15}^3	x_{16}^3	x_{17}^3	x_{18}^1	x_{19}^1	x_{20}^2	x_{21}^2	x_{22}^2	x_{23}^3	x_{24}^3	x_{25}^3	**N**	**Av**	
A_1	1	1	1	1																								=1
A_2					1	1																				**I**		=1
A_3							1	1																				=1
C_1									1	1	1																	=1
C_2												1	1	1														=1
C_3															1	1	1											=1
F_1																		1	1									=1
F_2																				1	1	1						=1
F_3																							1	1	1			=1
T1a	1	1			1																							=1
T1c									1	1		1		1			1										**I**	=1
T1f																		1		1			1	1				=1
T2a		1	1	1		1		1																				=1
T2c									1		1		1		1	1												=1
T2f																			1	1	1			1				=1
T3a	1		1		1	1	1																					=1
T3c										1		1																=1
T3f																		1				1	1		1			=1

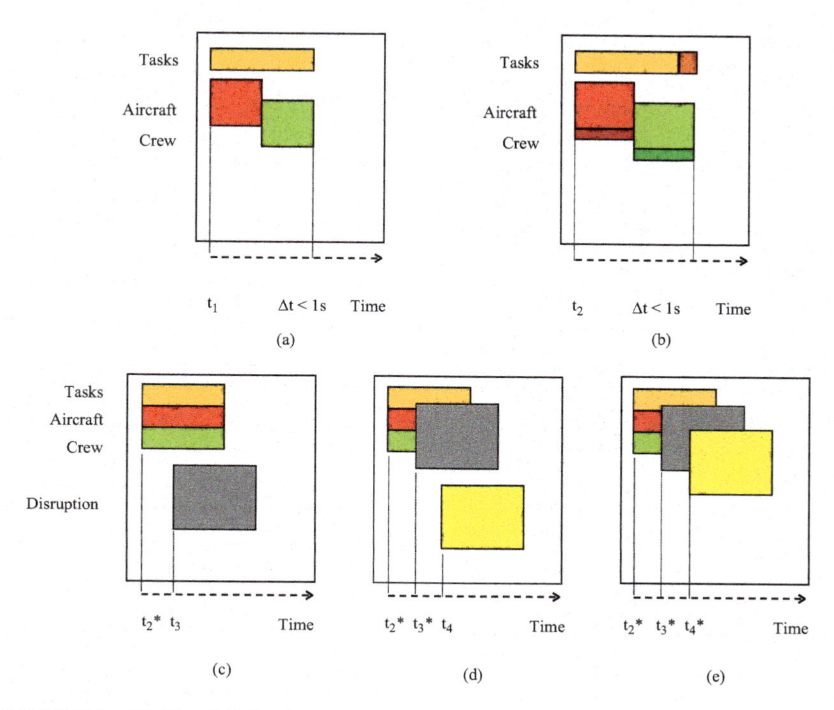

FIGURE 8.10 Rolling time horizon recovery approach.

Assume the airline schedule is operated as planned until the first disruption occurs at time t_1. As shown in Figure 8.10(a), the initial neighborhood is generated at time t_1 by including aircraft, crews, and tasks in it. The first problem is solved at t_1* (t_1* $=t_1+\Delta t$), where Δt is the problem solve time. If the first problem solved at t_1* is infeasible, then the neighborhood is expanded at t_2 (t_1* $< t_2$) by including more resources (Figure 8.10b) and the second problem is solved at t_2* (t_2* $= t_2+\Delta t$) as shown in Figure 8.10(c). At t_2*, the feasible solution of first disruption problem is obtained. However, as shown in Figure 8.10(d), second disruption occurs at t_3 (t_2* $< t_3$) and the second disruption scenario is solved at t_3* (Figure 8.10e). In case of more disruptions, say a third disruption which occurs at t_4 then the problem is solved at t_4* in the same manner as the previous ones (Figure 8.10e). All disruption problems are solved by taking the resources and tasks from the original schedule. Original schedule is the one which is operational before the disruption for which the problem is being solved. However, in case of the first disruption of the day, the planned schedule at the beginning of the day's operation is to be considered as the original schedule.

8.8 OTHER RECOVERY CONSIDERATIONS

In this chapter, the airline schedule recovery problem was addressed using the concept of disruption neighborhood generation. The problem is presented as a set partitioning problem to address the schedule recovery of crews for one aircraft fleet type to minimize the cost of resource allocation. However, at times, airline schedule recovery procedure may also demand optimal solutions to different recovery objectives, so as to minimize the cost of recovery procedure, minimize flight delay, minimize deviation from the planned schedule, minimize passenger inconvenience (broken itineraries), and so on. It is difficult to cater to all the objectives in one optimization problem and obtain a single solution which is optimal to all the aforementioned objectives. Therefore, in such scenarios, multi-objective optimization may be useful in dealing with complex operational issues. Problems with more than one objective generally have conflicting interests and they cannot be collectively optimized using a single objective function. A specific solution that provides optimal value for one objective may provide a compromised value for other objectives. Moreover, it is unlikely to find a single solution that offers optimal values to all the objectives. However, problems with more than one objective can be solved using multi-objective optimization approach. The optimization goal in such cases is to obtain a cluster of solutions providing different values of the objectives (often called a Pareto optimal set, which defines the Pareto frontier in the objective space) as shown in Figure 8.11. Pareto optimal solutions can be defined as those solutions which cannot be improved in any objective without necessarily worsening at least one other objective.

Pareto front can be obtained either by preference-based or by evolutionary algorithm-based multi-objective optimization. The basic distinction between these two approaches is that the number of optimization runs varies significantly in achieving the pareto front. Using the preference-based approach, the number of optimization runs would be significantly large, however similar results can be achieved in fewer optimization runs through evolutionary algorithm-based approach. A detailed description on multi-objective optimization approaches can be found in the works of Ehrgott (2005) and Deb (2005).

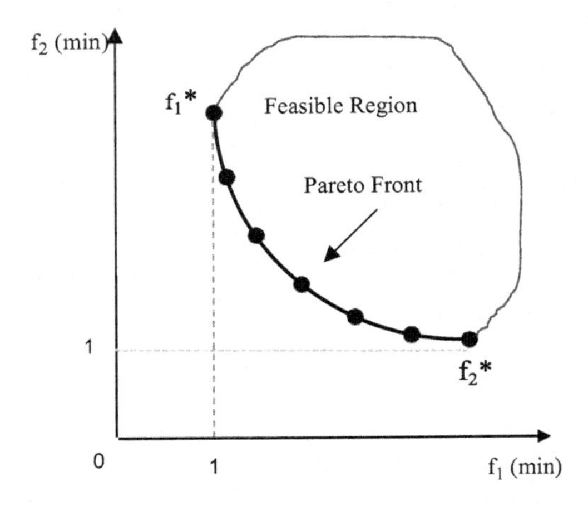

FIGURE 8.11 Pareto Front.

Most airlines operate with multiple fleet types to ensure profitability in their operations. However, from the recovery perspective it would be advantageous to have more than one fleet type involved in the recovery process to have swap opportunities among different fleet types. Swapping fleet types during schedule recovery can be considered if constraints such as appropriate crew type to operate an aircraft, destination port requirements with regards to an aircraft, and aircraft passenger-carrying capacity are satisfied. For instance, a crew cannot fly all fleet types and similarly all aircraft may not serve all the airports in the schedule. During recovery operations, swapping of fleet types needs to ensure that such constraints are satisfied. Figure 8.12 illustrates fleet swapping opportunity in a recovery neighborhood.

A recovery neighborhood shown in Figure 8.12 consists of a set of flights, aircraft (each from a different fleet type) and crews. Each flight F_n $\{n = 1, 2, ... 6\}$ is represented with respect to departure and arrival ports and times along with the resources, i.e., an aircraft A_i and crew C_i $\{i= 1, 2, 3\}$ involved with it.

Aircraft A_1 of flight F_1 is scheduled to do flight F_2 with crew C_1. Similarly, aircraft A_2 of flight F_3 is assigned to do flight F_4 with crew C_2 and aircraft A_3 of flight F_5 is to operate on flight F_6 with crew C_3. In this recovery neighborhood (Figure 8.12), there are swap opportunities (dashed arrow) for the resources at the port of disrutption (BOM). For instance, aircraft A_1 of flight F_1 can be swapped with aircraft A_3 of flight F_6 at BOM because crew C_3 of flight F_6 is eligible to fly aircraft A_1 of fleet type 1. A swap is possible with aircraft A_1 of flight F_2 and aircraft A_2 of flight F_3 because crew C_1 of flight F_2 can also fly aircraft A_2 of flight F_3 at BOM. Also, a swap can take place with aircraft A_2 of flight F_4 with aircraft A_3 of F_5 for the similar reason. However, aircraft A_1 of flight F_1 cannot operate on flight F_4 and aircraft A_3 of flight F_5 cannot operate on flight F_2 because of crew constraints. Moreover, aircraft A_3 of flight F_3 cannot operate on flight F_6 since destination of flight F_6 is not suitable for the fleet

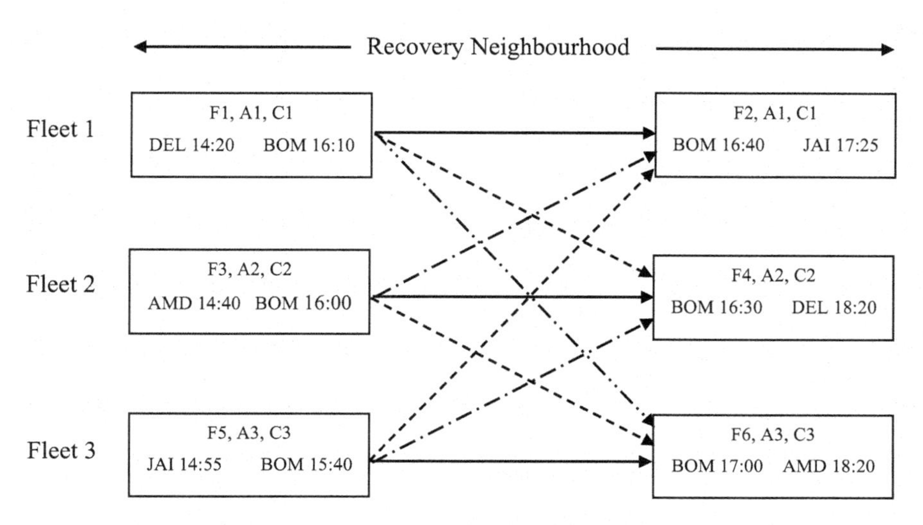

FIGURE 8.12 Swapping opportunities in the recovery neighborhood.

type of aircraft A_2 for operational constraints such as airport limitations to deal with large aircraft of a particular fleet type.

With the flexibility to swap the resouces from different fleet types, schedule recovery time window can be reduced significantly to resume to the originally planned schedule. The swapping of resources between differnet fleet types is a complex proposition especially due to union rules and industry regulations. At times, the legalities surrounding the swapping of resources can be a binding factor for the airline to prefer swapping over the delay option. Nevertheless, during schedule disruption, swapping of resources provides flexibility to the airline and saves them significant expenses and helps minimize incovenience to the passengers.

8.9 CONCLUSION

The chapter presents a unique insight into disruption management which is given significant attention by industry practitioners. We explore a disruption scenario wherein an airline schedule recovery problem is created by a delayed flight which causes disruption in the schedule. Despite rigorous planning and scheduling, on the day of operations, schedules don't always proceed according to plan owing to a number of unforeseen disruptions which result in a non-operational schedule. Once such a situation arises, schedule recovery is necessitated to return to the originally planned schedule at the earliest through the swapping of aircraft and crew and/or delaying or canceling of flights. Schedule recovery is achieved by the practice of neighborhood generation. During disruption the schedule is further hampered by broken aircraft routings and crew pairings which may result in extra costs incurred by the airline, operational problems at the airport, and inconvenience to passengers. There are a number of recovery approaches at the airline's disposal to recover aircraft, crew, and passenger itineraries. In disruption management, aircraft recovery is considered most critical, and is followed by crew recovery. However, of late, an integrated recovery of resources as well as passengers is also gaining attention within the industry. The advancements in computing capabilities and the development of optimization algorithms have substantially aided in disruption management.

CHAPTER QUESTIONS

Q1. Based on Figure 8.13, disruption in the schedule is observed at 12:00 h at AMD. How does the disruption in the schedule affect the aircraft routings?

Q2. Based on the schedule disruption presented in Figure 8.13, access the impact of broken crew pairings in the network?

Q3. Identify the affected resources and flights from the disruption scenario presented in Figures 8.13 and 8.14.

Q4. Generate the networks for aircraft and crews to cover the uncovered flights in the neighborhood.

Q5. Present the set partitioning representation of the network containing aircraft, crews, and the uncovered flights.

Q6. How does rolling time horizon approach work during multiple disruptions?

Q7. Discuss schedule recovery considerations to minimize the schedule recovery duration.

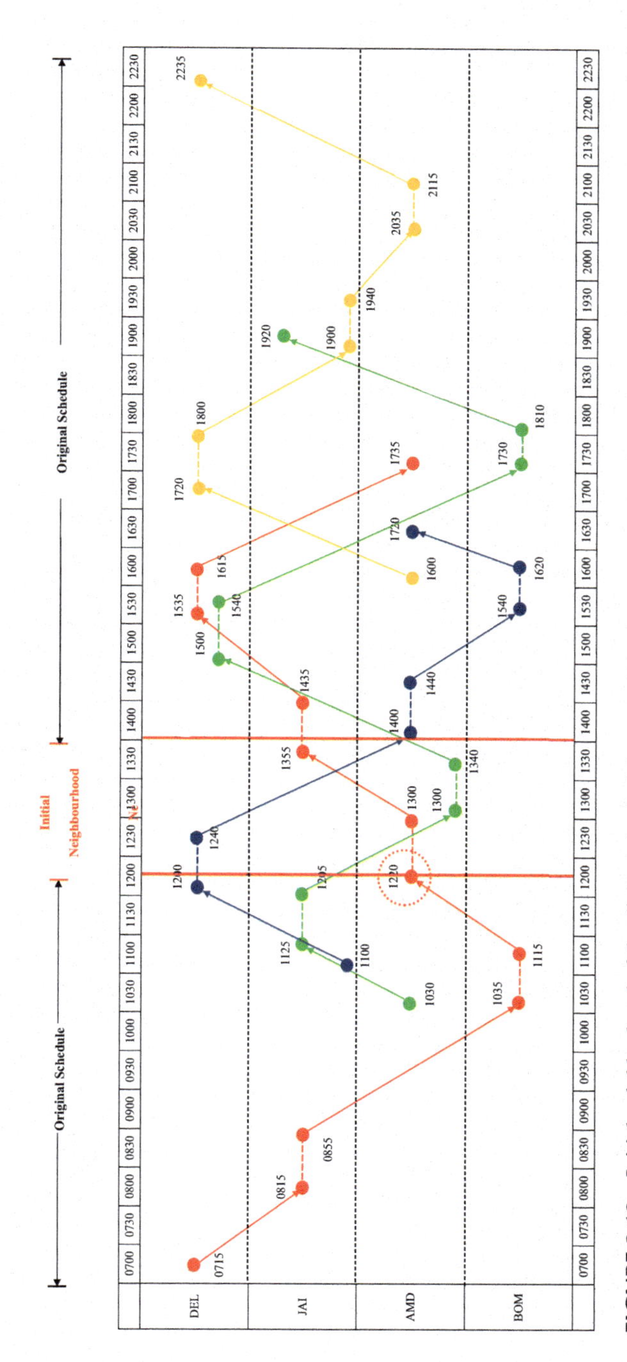

FIGURE 8.13 Initial neighborhood of the disrupted schedule.

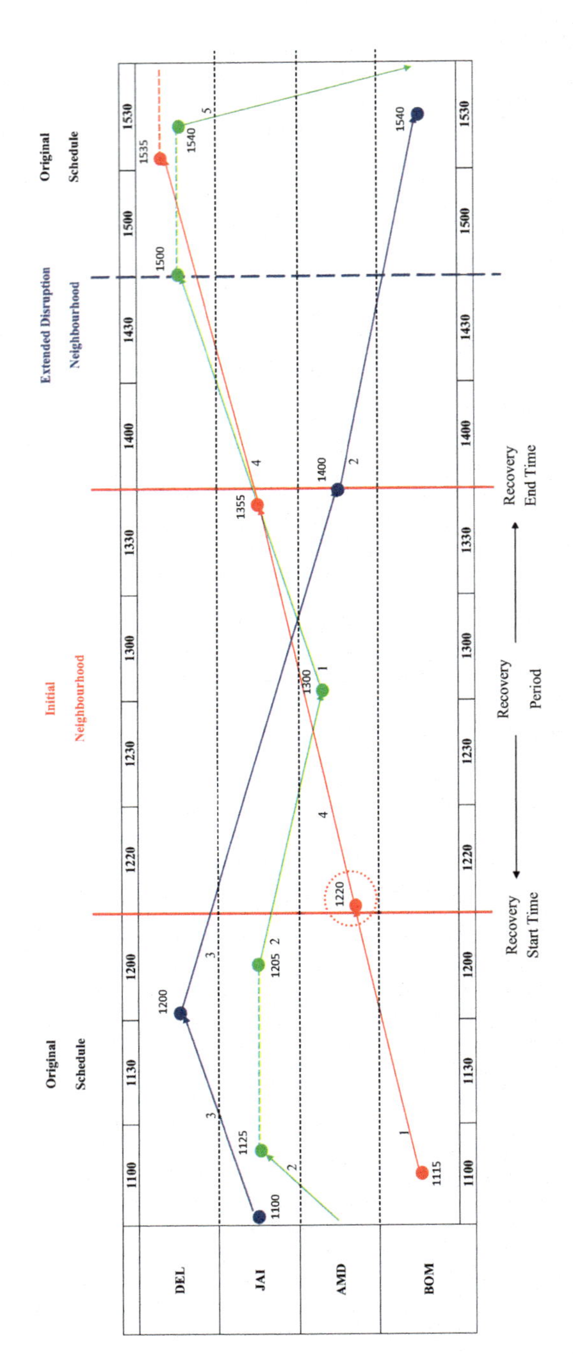

FIGURE 8.14 Extended neighborhood of the disrupted schedule.

REFERENCES

Deb, K. (2005). Multi-objective optimization. In E. K., Burke, G., Kendall (Eds), *Search Methodologies*. Springer.

Ehrgott, M. (2005). *Multicriteria Optimization*. Springer.

Guimarans, D., Arias, P., & Mota, M. M. (2015). Large neighbourhood search and simulation for disruption management in the airline industry. In Mujica Mota, M., De La Mota, I., Guimarans Serrano, D. (Eds), *Applied Simulation and Optimization*. Springer.

Ng, K. K. H., Keung, K. L., Lee, C. K. M., & Chow, Y. T. (2020). A large neighbourhood search approach to airline schedule disruption recovery problem, *IEEE International Conference on Industrial Engineering and Engineering Management (IEEM)*, pp. 600–604.

Rezanova, N. J., & Ryan, D. M. (2010). The train driver recovery problem – A set partitioning based model and solution method. *Computers and Operations Research*, *37*(5), 845–856.

9 Sustainability in the Aviation Industry

CHAPTER OBJECTIVES

At the end of this chapter, you will be able to

- Get an overview of the impact of aviation operations on the environment.
- Know the social and economic effects of the aviation industry.
- Understand waste management aspects at the airports.
- Know about sustainability reporting in the aviation industry.
- Identify the challenges in achieving sustainable aviation operations.
- Understand how sustainability can be achieved in the aviation industry.

9.1 INTRODUCTION

Since the groundbreaking flight of the Wright brothers in 1903, and over a century of operations, the environmental impact of the aviation industry has worsened and is finally being acknowledged globally. Now, besides safety aspects, sustainable operations are another area of significant concern for all parties. This chapter captures the practical approaches as well as policy considerations surrounding sustainability within the aviation industry. Sustainability was defined by the Brundtland Commission Report (1987) as 'meeting the needs of the present without compromising the ability of future generations to meet their own needs'. To achieve aviation sustainability, economic prosperity, social responsibility, and environmental stewardship, which is known as the Triple Bottom Line (TBL) approach, is considered (Landrum et al., 2012). Within the context of aviation operations, sustainability is primarily concerned with addressing the environmental impact of using fossil fuel and emissions of greenhouse gases (GHG), aircraft noise, and aviation waste (ICAO, 2013; Dessens et al., 2014). Additionally, aspects relating to human wellbeing, such as respiratory and cardiovascular conditions which result in mortality for those being affected by aviation operations, are also an area of concern. Owing to the nature, scope, and scale of the aviation industry, the environmental, social, and economic dimensions of sustainability affect a range of stakeholders both within and outside the industry.

DOI: 10.1201/9780203731338-9

It was around the late 1990s (Sledsens, 1998) when the concept of sustainability surfaced within the context of aviation operations. Sustainable aviation operations may simply be defined as people using air travel without compromising on natural resources or affecting the environment (Sledsens, 1998). However, from a technical standpoint, sustainability may be described as a state of systems having stable values for a given set of parameters within permissible bounds (Altuntas et al., 2019). In this context, a sustainable operation may be the ability of the aircraft to maintain its course without deviating from its flight path, unless intervened by the pilot. Therefore, for sustainable development in the aviation industry all its aspects, both on-ground and airborne, need to be considered (Altuntas et al., 2019). To achieve that, it's critical to know the specifications, consumption requirements, and volume of the resources used in aviation operations. In this chapter, the TBL approach is considered to get an overview of the sustainability dimensions of the aviation industry.

9.2 ENVIRONMENTAL SUSTAINABILITY

9.2.1 Air Pollution

Air pollution is the result of emissions including carbon monoxide (CO) and oxides of nitrogen (NOX) and sulfur (SOX) which originates from different operations at the airport such as refueling, ground transport vehicles, de-icing operations during winters etc. Similarly, aircraft-related pollution occurs based on the use of aircraft thrust, weather conditions, engine type, flight duration, altitude of the flight, state of tires and brakes, fuel efficiency etc. (Daley & Thomas, 2011). Manufacturers such as Rolls Royce, Pratt & Whitney, General Electric, GMF International etc. invest considerably on the development and improvement of aircraft engines to address emission concerns from the industry stakeholders (ICAO Aircraft Emissions, n.d.). Regardless, air quality is still significantly affected by aviation operations leading to local environmental issues. Therefore, for sustainable operations, the causes of air pollution need to be identified and addressed at various levels of the aviation operations.

Globally, the greenhouse gas emissions produced by the civil aviation sector ranges between 2% and 3% (Mayor & Tol, 2010; Owen et al., 2010) and with the growth in the international air travel and increase in the passenger movement, GHG emissions are unlikely to diminish in future of their own accord. Therefore, the current state of aviation operations is not sustainable ecologically (Chapman, 2007). Concerning the global impact of greenhouse gas emissions and its impact on the environment there is an ongoing debate in the airline industry and beyond to address the issues surrounding GHG emissions. During short-haul and long-haul flights, commercial aircraft fly at high altitudes ranging between 26,000 feet and 40,000 feet (Martine & Alves, 2015) and at these altitudes the emissions and other chemical residue such as carbon dioxide (CO_2), nitrogen oxides, and other pollutants and particulates released by the aircraft affect the atmosphere and contribute to the environmental imbalance (Yim et al., 2015; Budd et al., 2020). For instance, aircraft affect the climate as they leave a trail of cirrus clouds which stem from the emission of aerosol gases (Stordal et al., 2005). Emissions released at high altitudes have more adverse impact on the environment as opposed to those being released at sea or ground level (Budd et al.,

TABLE 9.1
Carbon emissions for around the globe (DEL–DXB–JFK–NRT–DEL) trip

Departure Airport	Arrival Airport	Distance (Km)	Aircraft Type	Aircraft Fuel Burn/ Leg (Kg)	Passenger CO2/Pax/ Leg (Kg)
DEL	DXB	2182	320	17,287	149.8
DXB	JFK	10,995	388	172,854	623.5
JFK	NRT	10,828	773	101,203	469.4
NRT	DEL	5906	788	37,438	346.6

2020). Therefore, there is a pressing need to mitigate the growing impact of aviation industry operations on the environment by establishing and imposing stringent standards through involving the relevant industry stakeholders. These environmental concerns are rather evident and fairly visible as opposed to other not-so-visible effects of the aviation operations. For instance, air transportation is also responsible as a causal factor in climate change, leading to reduced air quality (Harrison et al., 2015), water and land pollution (Daley & Thomas, 2011), and hazardous waste (Hooper & Greenall, 2005). Furthermore, there is a dearth of successful mechanisms to control GHG emissions due to conflicting objectives with the economic aspects. Also, traditionally, the approach has been to mitigate the emissions which is one of the reasons for the precedence of economic aspects. Contrarily, based on the United Nations Framework Convention on Climate Change, the paradigm shift in the approach can be to design low carbon emissions (not reliant on fossil fuel) into each aspect of the existing aviation processes (McManners, 2016).

To estimate emissions from air travel, ICAO developed an emissions calculator (ICAO tool, n.d.) based on aircraft type, travel distance, passenger load data, and so on. Through this, passengers can get an estimation of the CO2 emissions associated with their flight leg. Based on this, an estimate of a round the world air travel from Delhi (DEL) to Tokyo (NRT) via Dubai (DXB) and New York (JFK) is presented in Table 9.1.

As evident from Table 9.1, for a DEL–DXB–JFK–NRT–DEL trip covering 29,911 km, the fuel consumption is 328,782 kg. In this trip, CO2 emissions were 1589.3 kg per passenger. This is an indicator of the overall energy requirements and emissions from aviation operations. Therefore, industry stakeholders such as international civil aviation organizations, policy makers at international and local levels, governments, and the research community need to address the challenges posed by the aviation industry from a multitude of perspectives (De-Castro et al., 2016).

9.2.2 Noise Pollution

Apart from emissions, the aviation industry also impacts environmental sustainability through aircraft noise which has a substantial impact on the local environment (Postorino & Mantecchini, 2016). Aviation noise is considered as one of the most challenging and disliked aspects of the aviation operations among the communities living around airports all over the world. Aviation-related noise can be classified based on acoustic and non-acoustic factors. Acoustic factors include sound level, frequency, duration, seasonal and meteorological conditions, flight route etc. whereas non-acoustic aspects are noise sensitivity, fear of noise source, perceived predictability, control and coping mechanisms, noise compensation etc. (Zaporozhets & Blyukher, 2019).

For an aircraft, the source of noise is primarily generated from its engine and auxiliary power units. However, at the airport, the degree of noise varies and is subject to the equipment type, operating conditions, timing of the day, and number of takeoffs/landings (Daley & Thomas, 2011). Even though, over the years aircraft design has improved significantly (Appendix C) but aircraft noise is still considered as one of the limiting factors in sustainable aviation operations. Therefore, International Civil Aviation Organization (ICAO) monitors the advancements in noise reduction technology on a continuous basis and reviews the standards periodically for noise certification. According to Annex 16, volume 1 (ICAO, 2017), ICAO established noise certification to 'ensure that the latest available noise reduction technology is incorporated into aircraft design demonstrated by procedures which are relevant to day-to-day operations, to ensure that noise reduction offered by technology is reflected in reductions around airports'. It is important to know that the noise standards specifically apply to a particular aircraft type, i.e., the aeroplane.

Based on human perception of aircraft noise, duration of noise and sound frequency, Effective Perceived Noise in Decibels is used for noise certification purposes (Hardeman, 2020).

It is critical to understand that response to the noise generated from aviation operations is relevant with the expectations, attitudes, and perceptions of the communities that are affected by the noise. With constant exposure to aircraft noise, the communities living in the vicinity of the airports encounter a range of cognitive and physiological problems due to sleep deprivation and stress and suffer communication difficulties affecting their day-to-day functionalities (Hume & Watson, 2003; McManners, 2016). Therefore, modern-day airports are preferably located outside the periphery of the cities to minimize the effects of aviation operations on local communities. Nevertheless, apart from the toxic greenhouse emissions, aircraft noise constrains the aviation industry to address sustainability concerns (Gualandi & Mantecchini, 2008; Upham et al., 2003; Becken & Mackey, 2017). To address the concerns surrounding the assessment and control of aircraft noise in the vicinity of the airports, analytical modeling approaches are used to explain the relationships among the causes, risks, and impact of aircraft noise in a range of scenarios (Zaporozhets & Blyukher, 2019). For instance, low-flying, departing and arriving aircraft lead to significant noise on the ground especially around the hub airports (De Neufville & Odoni, 2013).

9.2.3 Aviation Fuel

Air transportation has witnessed continuous and global growth since the 1950s. Especially, in the last few decades, the annual air passenger traffic has increased considerably. Due to this, an increase in fuel and energy consumption is observed, resulting in higher greenhouse gas emissions. With the growth of the civil aviation industry, ever-increasing passenger volumes, and cargo shipments led to the stakeholders' concerns regarding the aviation industry's impact on the environment and energy consumption. For instance, between 2005 and 2015, energy consumption in the aviation industry has increased by more than 5% whereas in the same period the increase in the energy production has been less than 5% (Cui & Li, 2015). Unless suitable measures are taken at the appropriate levels by industry stakeholders, the gap between energy consumption and production is expected to increase further.

Environmental sustainability also focuses on establishing and maintaining a balance between renewable and non-renewable resources. Renewable fuel is defined as the fuel which is obtained from non-oil derivatives (electricity, liquid gas) and renewable sources (plants, animal waste). Fuel generated from renewable sources is conducive for the environment (Ekici et al., 2020). With the increasing passenger traffic in commercial aviation and an increase in the shipment of goods using air cargo services, jet fuel (high carbon emissions) demand is on the rise. Derivatives of jet fuel have been used but their adverse effect on the environment restricts their utility in combating environmental sustainability considerations. For economically feasible and environment-friendly aviation operations, biofuels (low carbon emissions) are being considered as an alternate option to reduce the GHG emissions emanating from the aircraft. However, the quantity of biofuel needed for aviation industry operations is considerably significant as opposed to jet fuel and therefore the biofuel supply sources need to be generated to ensure uninterrupted supply to meet the demand at multiple levels of aviation operations. However, over the years, numerous attempts have been made to tackle the issue of emissions and to enhance the operational efficiency through technological advancements made in the engine (turbine) and aircraft (aerodynamics) design and by using the scarce resources optimally.

It is important to know that according to international agreements (for international segments), aviation jet fuel is tax free (McManners, 2016) which makes it as a preferred fuel option for an airline despite its adverse impact on the environment. Even then, in aviation industry fuel costs are high and constitute the major portion of the expenditure to an airline's operations (McManners, 2016). The prices of jet fuel fluctuate and are based on the region and industry requirements. To give an overview of jet fuel prices, Table 9.2 presents jet fuel price and consumption data for the first week of May 2022 (IATA Jet Fuel, n.d.).

TABLE 9.2
Jet fuel prices

	Share in World Index	Cts/Gallon	$/Billion Barrel	$/Metric Ton
Jet Fuel Price	100%	419.74	176.29	1392.17
Asia & Oceania	22%	354.99	149.10	1177.86
Europe & CIS	28%	401.05	168.44	1329.00
Middle East & Africa	7%	366.21	153.81	1214.41
North America	39%	480.81	201.94	1595.31
Latin & Central America	4%	410.47	172.40	1361.93

9.2.4 Water Pollution

Similar to other industrial settings, at airports, water is essential for many activities such as cleaning and washing of aircraft, as a fire extinguisher, for the maintenance of infrastructure, to run day-to-day airport operations, and to cater to the customers' requirements who use airport's facilities as they arrive, depart, or use the airport as a transit point. For instance, the process to clean and wash the aircraft with water (using hazardous solvents) leaves toxic pollutants in the environment (Sulej et al., 2012). Similarly, aircraft de-icing and anti-icing solvents find their way into the water supply sources and pose threat to local ecology and human health (Sulej et al., 2012). Based on these activities which are carried out to ensure smooth functioning of aircraft and airport operations, water quality is impacted considerably causing contamination in the water bodies. As a result, there is a risk of compromising the health of the populace unless appropriate measures are taken to address the issues surrounding polluted surfaces and ground water which affect water supplies for drinking and agricultural purposes.

9.2.5 Climate Change

To manage environmental degradation due to aviation operations, the role and contribution of industry stakeholders is critical in establishing a sustainable aviation value chain (Weinhofer & Busch, 2013; Chiarini, 2014). For this, internal and external integration of different organizations of the value chain is required regarding sharing and managing of the resources to make decisions in addressing waste production, air and water pollution aspects that takes place across a range of aviation operations (Rizzi et al., 2013). Therefore, airlines and airports are required to constantly interact and engage at various levels with stakeholders by creating alliances to counter economic and environmental issues (Payán-Sánchez et al., 2017; Table 9.3).

To address global warming and climate change issues, International Civil Aviation Organization focuses on Carbon Offsetting and Reduction Scheme for International Aviation (CORSIA). Among other aspects, the aim of CORSIA is to stabilize aviation

TABLE 9.3
Stakeholder engagement and alliance regarding aviation environmental issues

	Stakeholder		
Issue	**Category**	**Engagement**	**Alliance**
Aircraft Noise	Communities, employees, passengers, airlines	High	Medium
Air pollution	Airlines, government, communities, airlines	High	Medium
Water pollution	Airports, passengers, authorities, suppliers	Medium	High
Waste production	Airports, airlines, suppliers, retailers, passenger	High	High
Climate change	Pressure groups, government, airlines, media	Low	Medium

industry's net CO_2 emissions. From 2020 onwards, CORSIA is expected to achieve carbon-neutral growth in the aviation sector and mitigate 2.5 billion tonnes of CO_2 before 2035 (CORSIA, n.d.).

9.3 SOCIAL AND ECONOMIC SUSTAINABILITY IN AVIATION

Besides material resources and environmental concerns, aviation industry impacts social aspects considerably since aviation operations revolve around human beings, therefore it needs to embrace social wellbeing. According to Al Sarrah et al. (2020), social sustainability can be addressed from the perspective of passengers (CSR policies, passenger rights, service quality), government authorities (noise levels, human rights, aviation quality control), airlines (crew health and safety, job opportunities, labor rights), and airports (passenger safety, cultural diversity, governance). As a result, social sustainability is another dimension in the aviation industry which is directly affected with the increased passenger demand. For instance, the infrastructure development (terminals, runways) in the aviation industry is not proportional with the passenger growth (Upham & Mills, 2005). As a result, there are numerous logistical issues that have surfaced such as increased delayed flights, missed flight connections, mishandling of luggage which eventually lead to customer dissatisfaction (Wald et al., 2010). Nevertheless, over the years, civil aviation has evolved significantly and developed into a dependable, reliable, safer, and fast mode of transportation resulting in numerous social benefits. In the process, the spin-off of the improvements (safety and speed) in the aviation industry has directly impacted hospitality, healthcare, and tourism industries (Perovic, 2013). Similarly, education, postal services and humanitarian aid have also significantly benefitted from the advancements in the aviation industry (Reddy & Thomson, 2015). It is important to know that civil aviation sector employs diverse and large number of employees (Kemp & Vinke, 2012; Palmer, 2016) and according to IATA, globally more than 60 million jobs are based in the aviation industry in which 10 million jobs are directly related with air transportation (IATA, 2017). Despite numerous challenges surrounding social sustainability concerns arising from aviation operations, attempts have been made to address it through transdisciplinary approaches (Diedrich et al., 2011).

As indicated in numerous studies, in comparison to environmental and social sustainability, economic sustainability is the most significant and lucrative aspect of the aviation industry. This is evident from the economic growth and diversification of the aviation sector globally (Chapman, 2007). In fact, in the aviation supply chains, organizations such as component suppliers, equipment manufacturers, airports, and subsidiary firms managing airport logistics operations have benefited with the development of civil aviation sector and reported significant financial gains (Junior et al., 2018). Moreover, social and environmental sustainability aspects are coupled with economic sustainability of the aviation operations. However, the triadic link between the three sustainability dimensions is dependent on the availability of the scarce natural resources (fuel) and the impact (adverse and favorable) of the development of the aviation industry on the society (ICAO, 2019).

9.4 SUSTAINABILITY IN AIRPORT OPERATIONS

Airports are an integral and significant part of the aviation industry and are responsible for 5% of aviation-induced CO2 emissions (ACI, n.d.). Due to socio-economic and environmental impacts intrinsic to aviation operations and in response to the pressure imposed by industry stakeholders (Upham & Mills, 2005), sustainable development of the airports is being considered globally. This is addressed by embedding sustainability considerations in the planning and design phases and in the operational aspects of the airports. With the current growth in the aviation industry, in many instances, airports are operational 24 h a day. Also, with the inclusion of new sectors in the airline flight networks, the number of airports has also been increased globally. Airports are the main drivers of aircraft noise, air, and water pollution and fuel related environmental concerns. Therefore, addressing sustainability concerns at the airport is important and demands significant attention from the industry stakeholders.

Waste management is another critical and major aspect in the airport operations and has a significant impact on the environment (Cowper-Smith & de Grosbois, 2011; Karaman et al., 2018; Postorino & Mantecchini., 2016). The volume of waste generated is proportional to passenger traffic and other commercial activities at the airport. According to IATA, aviation industry is responsible for the transportation of more than 7% of the world's GDP (EEA, 2018) which generates significant amount of cargo waste. Also, with the considerable increase in the passenger volumes and diversification of airport activities, the type and quantities of waste generated at the airports have increased manifold in the last couple of decades and have resulted in a new set of challenges. For instance, modern-day airports also include retail outlets, restaurants, recreational hubs for passengers leading to new streams of waste from the ground operations (Ferrulli, 2016). Retail and restaurant business operations alone generate more than 40% of the waste at the airports (Sebastian & Louis, 2021). Similarly, airports and airlines also produce significant amount of waste from maintenance activities, office, and terminal operations in the form of fuel spills, electronic waste, and pruning waste (Payán-Sánchez et al., 2017). Also, waste is incurred when the flight is airborne from inflight and cabin services (lavatory waste, food leftovers) which is unloaded at the arrival airport. However, waste generation in flights varies and depends on flight duration, i.e., there is a substantial difference in the volume of waste generated in the short- and long-haul flights. Waste mostly exists in the form of plastic, glass, metal, paper, rubber, and packaging material collected from a range of waste collection points at the airport. Therefore, to address waste-related sustainability aspects, innovative and state-of-the-art waste management practices need to be adopted, especially to counter post-COVID-19 changes. Airports are the major entry points to destinations across the globe and strict protocols for passenger movement in place as well as on-site COVID-19 testing facilities have resulted in biomedical wastes at the airports (Sebastian & Louis, 2021). Due to the global impact of COVID-19 and its subsequent variants, waste in the form of disposable masks (currently mandatory for air travel) has emerged as a significant source of microplastics in the environment (Fadare & Okoffo, 2020). For instance, in India, the major airports, i.e., Delhi, Mumbai, Chennai, Hyderabad, Kolkata, and Bengaluru generate around six tonnes of biowaste every day which was non-existent before COVID-19. Moreover, different

TABLE 9.4
Waste management at airports

Airport Waste	Waste Source	Waste Type
Terminal	Public areas, Offices, Operational activities	Toxic substances, Oil spills, Cleaning, Wastewater
Hangers	Service Equipment	Batteries, Tires, Fuel oil Transmission fluids
Tenant Commissaries	Restaurants, Retail stores, Hotels	Food, Packaging, Paper, Plastic
Cargo	Public and Private entities	Pallets, Packaging,
Medical	Personal Protective Equipment (PPE)	Gloves, Masks, Shields
Construction Material	Maintenance, Development	Wood, Bricks, Asphalt, Gravel, Stones, Soil
Aircraft Waste	Waste Source	Waste Type
Catering	Passengers/Crews food and Beverage leftovers	Bottles, Cups, Towels, Wrappers, Paper
Cleaning	Passengers/Crews, Leftovers Washroom bins, Lavatory	Chemicals, Pathogens

airlines also use and dispose of more than 75,000 personal protective equipment on a daily basis (Chowdhury & Rajagopal, 2020). Sustainable use of energy for airport operations is a relatively little explored domain within the aviation industry. With the increase in passengers, inclusion of new destinations in the flight networks, increase in airport capacity, and construction of new airports results in the increased demand for energy sources. Globally, the demand for energy is largely fulfilled using the fossil fuel sources such as oil, gas, and in other instances, through coal. The use of fossil fuels to satisfy the energy needs is not sustainable as they are depleting rapidly and are thus non-renewable. Moreover, such energy sources contribute toward global warming significantly. Therefore, for sustainable operations, it is required to generate energy from renewable sources and for that to be done efficiently. In summary, Table 9.4 presents different types and sources of waste at airports.

At airports, more than 75% of the energy is consumed at the terminal building, whereas Radio Navigation Systems, airspace lighting, parking, fire station, and other operations combined tally up the rest of the energy consumption (Akyuz et al., 2019). At the terminal building, the most energy intensive systems are Heating, Ventilating and Air Conditioning systems which consume around 70% of the energy, while the remaining 30% is consumed by lighting and other terminal building operations at the airport (Akyuz et al., 2019).

9.5 SUSTAINABILITY REPORTING IN AVIATION INDUSTRY

The trend to capture sustainability aspects and publishing sustainability reports has caught on, such that studies have captured the environmental sustainability aspects in aviation industry operations (Koc & Durmaz, 2015). However, the lack of sustainability initiatives across airline and airport operations and the reluctance to publish sustainability reports has raised concern (Kılıç et al., 2019).

In 1997, United Nations Environmental Program and the Coalition of Environmentally Responsible Economies founded the Global Reporting Initiative (GRI) as an international non-profit organization (Laskar & Maji, 2016). Since its inception, GRI is recognized as a global standard for sustainability reporting. Initially, the focus of GRI was confined only to environmental issues, however, over a period, economic and social aspects were also included in the sustainability reporting framework to assess organizational performance (Jain & Winner, 2016). The GRI sustainability framework is widely accepted and recognized for corporate social reporting and many organizations from different industrial backgrounds use GRI guidelines for CSR (Nikolaeva & Bicho, 2011; Martínez-Ferrero et al., 2013). One of the significant aspects of GRI is to ensure comparable sustainability reporting (temporally) across organizations (Isaksson & Steimle, 2009; Einwiller et al., 2016).

Sustainability reporting in the aviation industry is primarily focused on the environmental aspects through standalone reports capturing organizational practices and highlighting their efforts for environmental sustainability in the operations. Major airlines across the world acknowledge the significance of publishing environmental sustainability reports to comply with the requirements of aviation industry regulators (Mak & Chan, 2006) and to project themselves as a responsible member of the aviation community. To achieve this, airlines invest resources and effort to suit their business objectives. As compared to the emphasis on environmental concerns, the focus on social or economic aspects of CSR reporting in the aviation industry is comparatively insignificant (Cowper-Smith & de Grosbois, 2011). Similar to airlines, sustainability reporting of the airports is also subject to variation and focuses on environmental issues (Skouloudis et al., 2012). One of the reasons for the variation in the sustainability reporting of aviation industry can be attributed to the fact that airlines operate in different parts of the world and are placed in different regions that function under different regulations, legal frameworks and governance structures, have variation in the infrastructural development of the region and/or countries, and observe different cultural traditions which affect CSR reporting (Sheldon & Park, 2011). However, GRI mitigates the degree of subjectivity in the interpretation and understanding of sustainability reporting (Legendre & Coderre, 2013). Therefore, airlines adhere to the guidelines by adopting to the norms of the GRI framework and comply with the norms of the industry (Nikolaeva & Bicho, 2011). In sustainability reporting, the role and significance of regulatory bodies and institutions in the aviation industry is critical in influencing through providing incentives for adherence and penalizing for non-compliance of the standardized sustainability reporting (Yang & Rivers, 2009; Momin & Parker, 2013).

9.6 CHALLENGES IN AVIATION SUSTAINABILITY

Aviation industry is one of the more challenging industries to achieve sustainable operations, especially in minimizing emissions without compromising on the mobility aspects (Gössling & Upham, 2009). For global aviation operations, International Civil Aviation Organization (ICAO) establishes policies and sets standards with the aim 'to achieve sustainable growth of the global aviation system' (ICAO, 2016). This aim considers policy making to achieve the TBL approach through economic, social, and environmental sustainability and attempts to ensure a trade-off between operational efficiency and sustainability through stakeholder engagement (Al Sarrah et al., 2020). Similarly, there is growing realization among airline operators, airports, aircraft manufacturers, and industry regulators that the effect of aviation operations on the environment is a significant constraint on the development of the industry (Budd et al., 2020). Aviation industry is one of the best examples of addressing conflicting objectives between economic policy and environmental concerns. In fact, in aviation operations, the dilemma between achieving economic benefits and keeping the adverse impact of aviation operations within permissible bounds is not only challenging but also being more widely addressed (Walker & Cook, 2009). However, to establish the sustainability pathway, in the trade-off between these two aspects, economic objectives almost always dominate over environmental considerations (McManners, 2016). Conversely, one of the approaches is to consider environmental sustainability as the focus of operations and make it economically viable as opposed to considering economic solutions and making it environmentally sustainable (McManners, 2016).

In aviation industry, there is a growing recognition of achieving sustainability objectives (short-term) and goals (long-term); however, the mechanism to achieve sustainable aviation operations varies and largely remains unclear. Due to this, the major challenge is to strike a balance between the TBL aspects and other aviation industry considerations such as safety, timeliness, and accessibility. This leads to considerable divergence between practical aviation operations and institutional policies set out by local and international industry actors (Barr, 2012). Nevertheless, the environmental impact of the aviation operations is the focus of sustainability policy in the aviation sector which seldom emphasizes on economic or social dimensions (Monsalud et al., 2014).

9.7 ACHIEVING SUSTAINABILITY IN AVIATION INDUSTRY

As discussed earlier, aviation is a significant driver for trade, employment, movement, and economic growth. Consequently, it is important not only to ensure safe air travel and security but also to mitigate the effect of aviation operations on the environment through technological improvements using less carbon intensive fuel options, through environment-friendly policy and managing waste, as presented below.

9.7.1 Technological Improvements

In aviation industry, significant emphasis is placed on the research and development of high-end technology. Therefore, to achieve sustainability in aviation industry operations, the use of technology is considered as one of the important ways to move forward in achieving sustainability goals. For this, technological advancements in aircraft design (aerodynamics, propulsion, weight and structure, and control), air traffic management, engine design, use of alternative non-fossil fuels, and state-of-the-art navigational tools are required (Lutte & Bartle, 2017). Capturing these aspects, other flying options can be generated for commercial and cargo aviation. For instance, a hybrid vehicle (relatively slow in speed but with less emissions) with the capabilities of an airship and aircraft can be considered for short-haul flights in shipping large number of passengers (McManners, 2016). Few attempts are being made to develop hybrid flying options for commercial use (Khoury, 2012; Ceruti & Marzocca, 2014) by considering laminar flow control, blended wind body configuration, and gull-boxed wings (Ahmad & Kurtulus, 2019). However, the demand for such services is expected to cater to niche segments with the major operations continuing to focus on the existing aviation scenario (high speed but with more emissions) but the growth in passenger travel for short-haul flights may be effective to significantly trade-off between speed and carbon emissions. IATA also places considerable emphasis on the use of improvised technology to mitigate GHG emissions (Rotger, 2012). Using sophisticated real-time navigational algorithms can be effective to generate optimal flight paths in reducing carbon emissions, minimizing operating costs, and ensuring scheduled flight operations.

9.7.2 Fuel Options

With the acknowledgment of the adverse effect of jet fuel (mostly kerosene; non-renewable) on the environment, alternative and renewable fuel options are being explored to make the shift from fossil fuels in order to mitigate carbon emissions (European Aviation Environmental Report, 2016). As the term alternative fuel suggests, it is defined as the fuel which is used as an alternative to diesel and traditional gasoline (Ekici et al., 2020). Alternative fuels such as natural gas, hydrogen, butanol, and vegetable oils are derived from diverse resources and are more environment friendly. Therefore, there is an increasing consideration in the aviation industry to use alternate fuels and renewable fuel to mitigate the environmental impact of the conventional jet fuel.

Moreover, for international aviation operations imposing fuel tax and carbon tax is another way to strike out a balance between economic and environmental sustainability. For instance, in conventional aircraft (high speed, more emissions/pax) operations are expensive whereas the use of hybrid vehicles (low speed, less emissions/pax) could be an option for short-haul flight. For this, the customers who are not much aware of the impact of civil aviation operations on the environment will need to be educated as their support is paramount to successfully establishing sustainable aviation operations (Gössling & Peeters, 2007; Lassen, 2010). Other options to operate aircraft include, but are not limited to, solar power, hydrogen fuel cell, and liquid hydrogen (Ahmad & Kurtulus, 2019).

9.7.3 Aviation Policy

Sustainable operations largely depend on the decision makers who comply with the policies devised to run various aviation functions. The decisions are made on the short-term, mid-term, and long-term basis to implement operational strategies where it is critical to consider real-world challenges and stakeholder obligations. In many instances, there is a saturation point in aviation operations where economic and environmental policy aspects need to be evaluated and modified to achieve sustainability objectives of modern-day airlines. Furthermore, it is required to examine both at the planning and operational levels how regulations are practiced in the industry. With the lack of consensus on sustainability standards across the aviation industry, it is imperative for the policy makers to establish an agreed upon understanding of the meaning of sustainable operations and meet those sustainability standards using a globally recognized certification mechanism (Palmer, 2020).

The standards can be established from the perspective of legalities, planning, monitoring and continuous improvement, GHG emissions, human and labor rights, rural and social development, local food security, conservation, soil, water, air, use of technology, management of waste, and land rights (Palmer, 2020). Through this, compliance with the environmental regulations can be established, impact on climate can be mitigated, noise pollution can be reduced, local communities can be protected, and energy and resource conservation and waste management aspects can be addressed (Radomska & Cherniak, 2020). Therefore, it is imperative for policy makers in the aviation industry to accommodate these considerations to capture a comprehensive sustainability framework. In such a framework, economic efficiency, optimal use of natural (renewable and non-renewable) and artificial (terminals, hangers, runways) resources and protection of environment (the ozone layer, biodiversity, global climate) forms the fundamental basis of interaction between the industry stakeholders (Radomska & Cherniak, 2020). Similar to other aspects, noise and CO2 have considerably different impact on the environment as their area of influence varies in time and scale. Due to these concerns, ICAO developed standards in various domains to standardize the practices for the sustainable development of the aviation industry and to mitigate the impact of emissions on the environment (Cui & Li, 2015).

9.7.4 Waste Management Aspects

Waste management is another area of significance where sustainability aspects are required to be addressed. There is a considerable volume of waste in the form of wood, plastic, and paper (Hooper & Greenall, 2005) generated at the airports from civil aviation and cargo operations which can be recycled, put to reuse or reduced with proper usage. More than 75% of the waste generated at the airports is subject to recycling or can be used as a compost. However, due to inappropriate waste management practices the waste is not classified as recyclable, reusable, or non-cyclable waste and ends up in landfills (Li et al., 2003; Mehta, 2015; Sebastian & Louis, 2021). Therefore, waste not managed with due sustainability considerations in the aviation supply chain poses a threat not only to the environment through pollutants and contaminations but also to human wellbeing, flora, and fauna (Sebastian & Louis, 2021).

Lean approach in the aviation operations is imperative to address waste concerns across the range of airport and airline operations. However, incorporating lean methodology in the service industry operations is challenging due to the subjectivity involved across its implementation, interpretation, and aspects related to change management. Moreover, the lack of understanding surrounding the lean philosophy further limits its application in the service industry since lean approaches are predominantly ascribed to the manufacturing sector (Syltevik et al., 2018).

9.8 CONCLUSION

The aviation industry has always been faced by the challenge of balancing between enhancing the economic gains for the airlines and benefiting society through minimizing the impact of its operations on the environment. And while the effects of the aviation industry on the environment are being addressed, its impact on the socio-economic aspects remains contentious. Regardless, sustainability issues surrounding the aviation industry affect stakeholders on various levels: locally, regionally, and globally (Schäfer & Waitz, 2014). The large and diversified workforce demands social and economic sustainability from airline operators, which is to say that with growth and expansion in aviation operations, the consideration for economic and social sustainability is also being necessitated. Resultantly, the adverse impact of operations is not confined to specific areas and extends beyond localized boundaries. While the degree of impact varies, the vulnerable regions bear the brunt of the subjective and, at times, not specific regulations and legal mechanisms placed to curb the impact of the aviation industry.

CHAPTER QUESTIONS

Q1. In the context of airline operations, discuss the impact on environmental, social and economic sustainability.

Q2. Discuss the impact of airport operations on environmental, social, and economic sustainability.

Q3. How can waste management be improved at the airports? Discuss.

Q4. Identify two major areas for improvement in airline and airport operations based on the sustainability initiatives taken by aviation regulatory bodies.

Q5. Discuss the trade-off between the renewable and non-renewable fuel options in the aviation industry.

REFERENCES

ACI. (2011). Airport carbon accreditation annual report. www.airportcarbonaccreditation.org/about/co2reduction.html

Agrawal, A. (2020). Sustainability of airlines in India with Covid-19: Challenges ahead and possible way-outs. *Journal of Revenue and Pricing Management*, *20*, 457–472.

Ahmad, T., & Kurtulus, D. F. (2019). Technology review of sustainable aircraft design. In T. H. Karakoç, C. O. Colpan, O. Altuntas & Y. Sohret (Eds), *Sustainable Aviation* (pp. 141–146). Springer.

Akyuz, M. K., Altuntas, O., Sogut, M. Z., & Karakoç, T. H. (2019). Energy management at the airports. In T. H. Karakoç, C. O. Colpan, O. Altuntas & Y. Sohret (Eds), *Sustainable Aviation* (pp. 9–36). Springer.

Al Sarrah, M., Ajmal, M. M., & Mertzanis, C. (2020). Identification of sustainability indicators in the civil aviation sector in Dubai: A stakeholders' perspective. *Social Responsibility Journal*, *17*(5), 648–668.

Altuntas, O., Sohret, Y., & Karakoç, T. H. (2019). Fundamentals of sustainability. In T. H. Karakoç, C. O. Colpan, O. Altuntas & Y. Sohret (Eds), *Sustainable Aviation* (pp. 3–5). Springer.Barr, S. (2012). *Environment and society: Sustainability policy and the citizen*. Ashgate.

Becken, S. & Mackey, B. (2017). What role for offsetting aviation greenhouse gas emissions in a deepcut carbon world? *Journal of Air Transport Management, 63*, 71–83.

Brundtland Commission Repor.t (1987). Our Common Future.

Budd, T., Intini, M., & Volta, N. (2020). Environmentally sustainable air transport: A focus on airline productivity. In T. Walker, A. S. Bergantino, N. S. Much, L. Loiacono (Eds), *Sustainable Aviation: Greening the Flight Path* (pp. 55–77). Palgrave Macmillan.

Ceruti, A., & Marzocca, P. (2014). Conceptual approach to unconventional airship design and synthesis. *Journal of Aerospace Engineering, 27*(6). https://trid.trb.org/view/1309890

Chapman, L. (2007). Transport and climate change: A review. *Journal of Transport Geography, 15*(5), 354–367.Chiarini, A. (2014). Strategies for developing an environmentally sustainable supply chain: Differences between manufacturing and service sectors. *Business Strategy and the Environment, 23*(7), 493–504.

Chowdhury, A., & Rajagopal, D. (2020). *Covid-19 heaps up bio waste at airports*. https://economictimes.indiatimes.com/industry/transportation/airlines-/-aviation/covid-19-heaps-up-bio-waste-at-airports/articleshow/79552264.cms?from=mdr

CORSIA. (n.d.). CORSIA explained https://aviationbenefits.org/environmental-efficiency/climate-action/offsetting-emissions-corsia/corsia/corsia-explained/

Cowper-Smith, A., & de Grosbois, D. (2011). The adoption of corporate social responsibility practices in the airline industry. *Journal of Sustainable Tourism, 19*(1), 59–77.

Cui, Q., & Li, Y. (2015). Evaluating energy efficiency for airlines: An application of VFB-DEA. *Journal of Air Transport Management, 44*, 34–41.

Daley, B., & Thomas, C. (2011). Challenges to growth: Environmental issues and the development of the air transport industry. In J. F. O'Connell & G. Williams (Eds), *Air Transport in the 21st Century* (pp. 269–294). Ashgate Publishing.

De-Castro, M. G., Amores-Salvadó, J., & Navas-López, J. E. (2016). Environmental management systems and firm performance: Improving firm environmental policy through stakeholder engagement. *Corporate Social Responsibility and Environmental Management, 23*(4), 243–256.

De Neufville, R., Odoni, A. R., Belobaba, P. P. & Reynolds, T. G. (2013). Airport Systems: Planning, Design, and Management (2 ed.), McGraw-Hill.

Dessens, O., Köhler, M. O., Rogers, H. L., Jones, R. L., & Pyle, J. A. (2014). Aviation and climate change. *Transport Policy, 34*, 14–20.

Diedrich, A., Upham, P., Levidow, L., & van den Hove, S. (2011). Framing environmental sustainability challenges for research and innovation in European policy agendas. *Environmental Science & Policy, 14*(8), 935–939.

Einwiller, S., Ruppel, C. & Schnauber, A. (2016). Harmonization and differences in CSR reporting of US and German companies: Analyzing the role of global reporting standards and country of origin. *Corporate Communications: An International Journal, 21*(2), 230–245.

Ekici, S., Orhan, İ., Karakoç, T.H., & Hepbasli, A. (2020). Milestone of greening the flight path: Alternative fuels. In T. Walker, A. S. Bergantino, N. S. Much, L. Loiacono (Eds), *Sustainable Aviation: Greening the Flight Path* (pp. 243–253). Palgrave Macmillan.

European Aviation Environmental Report. (2016). https://ec.europa.eu/transport/sites/transport/files/european-aviation-environmental-report-2016-72dpi.pdf

European Environment Agency. (2020). *Annual European union greenhouse gas inventory 1990–2018 and inventory report 2020*. www.eea.europa.eu//publications/european-union-greenhouse-gas-inventory-2020

Fadare, O. O., & Okoffo, E. D. (2020). Covid-19 face masks: A potential source of microplastic fibers in the environment. *Science of the Total Environmental, 737*, 140279.

Ferrulli P. (2016). Green airport design evaluation (GrADE) – Methods and tools improving infrastructure planning. *Transportation Research Procedia*, *14*, 3781–3790.

Gössling, S., & Peeters, P. (2007). It does not harm the environment! An analysis of industry discourses on tourism, air travel and the environment. *Journal of Sustainable Tourism*, *15*(4), 402–417.

Gössling, S., & Upham, P. (2009). *Climate Change and Aviation: Issues Challenges and Solutions*. Earthscan.

Gualandi, N., & Mantecchini, L. (2008). Aircraft noise pollution: A model of interaction between airports and local communities. *International Journal of Mechanical Systems Science and Engineering*, *2*(2), 137–141.

Hardeman, A. B. (2020). Sustainable alternative air transport technologies. In T. Walker, A. S. Bergantino, N. S. Much, L. Loiacono (Eds), *Sustainable Aviation: Greening the Flight Path* (pp. 277–305). Palgrave Macmillan.

Harrison, R. M., Masiol, M., & Vardoulakis, S. (2015). Civil aviation, air pollution and human health. *Environmental Research Letters*, *10*(4), 1–3.

Hooper, P. D., & Greenall, A. (2005). Exploring the potential for environmental performance benchmarking in the airline sector. *Benchmarking: An International Journal*, *12*(2), 151–165.

Horonjeff, R., McKelvey, F. X., Sproule, W. J., Young, S. B. (2010). *Planning & Design of Airports*. McGraw Hill.

Hume, K., & Watson, A. (2003). The human health impacts of aviation. In P. Upman, J. Maughan, D. Raper, & C. Thomas (Eds), *Towards Sustainable Aviation*, Routledge.

International Air Transport Association. (2017). Aviation benefits. www.iata.org/policy/documents/aviation-benefits-%20web.pdf

International Air Transport Association. (n.d.). *Jet Fuel Price Monitor*. www.iata.org/en/publications/economics/fuel-monitor/

International Civil Aviation Organization. (2017). *Aircraft Noise*. https://store.icao.int/en/annex-16-environmental-protection-volume-i-aircraft-noise

International Civil Aviation Organization. (n.d.). *Aircraft Emissions Databank*. www.easa.europa.eu/domains/environment/icao-aircraft-engine-emissions-databank

International Civil Aviation Organization. (n.d.). *Carbon Emissions Calculator*. www.icao.int/environmental-protection/CarbonOffset/Pages/default.aspx

International Civil Aviation Organization Secretariat. (2013). *Reducing Aircraft Noise*. ICAO Environmental Report 2013, 56–57.

International Civil Aviation Organization. (2016). *ICAO vision and mission*. Retrieved from www.icao.int/about-icao/Pages/vision-and-mission.aspx

International Civil Aviation Organization. (2019). Global Aviation and Our Sustainable Future: International Civil Aviation Organization Briefing for RIO+20, Montreal.

Isaksson, R., & Steimle, U. (2009). What does GRI-reporting tell us about corporate sustainability? *The TQM Journal*, *21*(2), 168–181.

Jain, R., & Winner, L. H. (2016). CSR and sustainability reporting practices of top companies in India. *Corporate Communications: An International Journal*, *21*(1), 36–55.

Junior, B. A., Majid, M. A., & Romli, A. (2018). Green information technology for sustainability elicitation in government-based organisations: An exploratory case study. *International Journal of Sustainable Society*, *10*(1), 20–41.

Karaman, A. S., Kılıç, M., & Uyar, A. (2018). Sustainability reporting in the aviation industry: Worldwide evidence. *Sustainability Accounting, Management & Policy Journal*, *9*(4), 362–391.

Kemp, L. J., & Vinke, J. (2012). CSR reporting: A review of the Pakistani aviation industry. *South Asian Journal of Global Business Research*, *1*(2), 276–292.

Khoury, G. A. (2012). Airship technology. In *Cambridge Aerospace Series* (2nd Ed). Cambridge University Press.

Kılıç, M., Uyar, A., & Karaman, A. S. (2019). What impacts sustainability reporting in the global aviation industry? An institutional perspective. *Transport Policy, 79*, 54–65.

Koç S., & Durmaz, V. (2015). Airport corporate sustainability: An analysis of indicators reported in the sustainability practices. *Social and Behavioral Sciences, 181*, 158–170.

Landrum and Brown, Inc., Environmental Consulting Group, Inc., Primera Engineers, Ltd., & Muller and Muller, Ltd. (2012). *Guidebook for incorporating sustainability into traditional airport projects*. Airport Cooperative Research Program: Report 80, Transportation Research Board for the National Academies, Washington, DC.

Laskar, N., & Maji, S. G. (2016). Corporate sustainability reporting practices in India: Myth or reality? *Social Responsibility Journal, 12*(4), 625–641.

Lassen, C. (2010). Environmentalist in business class: An analysis of air travel and environmental attitude. *Transport Reviews, 30*(6), 733–751.

Legendre, S., & Coderre, F. (2013). Determinants of GRI G3 application levels: The case of the fortune global 500. *Corporate Social Responsibility and Environmental Management, 20*(3), 182–192.

Li, X. D., Poon, C. S., Lee, S. C., Chung, S. S., & Luk. F. (2003). Waste reduction and recycling strategies for the in-flight services in the airline industry. *Resources, Conservation and Recycling, 37*(2), 87–99.

Lutte, R. K., & Bartle, J. R. (2017). Sustainability in the air: The modernization of international air Navigation. *Public Works Management and Policy, 22*(4), 322–334.

Mak, B., & Chan, W. W. (2006). Environmental reporting of airlines in the Asia pacific region. *Journal of Sustainable Tourism, 14*(6), 618–628.Martine, G., & Alves, J. E. D. (2015). Economy, society and environment in the 21st century: Three pillars or trilemma of sustainability? Revista Brasileira de Estudos de População, *32*(3), 433–459.

Martínez-Ferrero, J., Garcia-Sanchez, I. M., & Cuadrado-Ballesteros, B. (2013). Effect of financial reporting quality on sustainability information disclosure. *Corporate Social Responsibility and Environmental Management, 22*(1), 45–64.

Mayor, K., & Tol, R. S. J. (2010). Scenarios of carbon dioxide emissions from aviation. *Global Environmental Change, 20*(1), 65–73.

Mehta, P. (2015). Aviation waste management: An insight. *International Journal of Environmental Sciences, 5*(6), 179–186.

McManners, P. J. (2016). Developing policy integrating sustainability: A case study into aviation. *Environmental Science & Policy, 57*, 86–92.

Momin, M. A., & Parker, L. D. (2013). Motivations for corporate social responsibility reporting by MNC subsidiaries in an emerging country: The case of Bangladesh. *The British Accounting Review, 45*(3), 215–228.

Monsalud, A., Ho, D., & Rakas, J. (2014). Greenhouse gas emissions mitigation strategies within the airport sustainability evaluation process. *Sustainable Cities and Society, 14*, 414–424.

Nikolaeva, R., & Bicho, M. (2011). The role of institutional and reputational factors in the voluntary adoption of corporate social responsibility reporting standards. *Journal of the Academy of Marketing Science, 39*(1), 136–157.

Owen, B., Lee, D. S., & Ling, L. (2010). Flying into the future: Aviation: Emissions scenarios to 2050. *Environmental Science and Technology, 44*(7), 2255–2260.Palmer, W. (2020). Sustaining flight: Comprehension, assessment, and certification of sustainability in aviation. In T. Walker, A. S. Bergantino, N. S. Much, L. Loiacono (Eds), *Sustainable Aviation: Greening the Flight Path* (pp. 6–28). Palgrave Macmillan.

Palmer, W. J. (2016). Will Sustainability Fly?: Aviation Fuel Options in a Low-Carbon World (pp. 43–46), Routledge.
Payán-Sánchez, B., Plaza-Úbeda, J. A., Pérez-Valls, M. & Carmona-Moreno, E. (2017). Social embeddedness for sustainability in the aviation sector. *Corporate Social Responsibility and Environmental Management*, *25*(4), 537–553.
Perovic, J. (2013). The economic benefits of aviation and performance in the travel & tourism competitiveness index. In *the Travel & Tourism Competitiveness Report 2013* (pp. 57–61), World Economic Forum.
Postorino, M. N., & Mantecchini, L. (2016). A systematic approach to assess the effectiveness of airport noise mitigation strategies. *Journal of Air Transport Management*, *50*, 71–82.
Radomska, M., & Cherniak, L. (2020). The analysis of the sustainability commitment formulation and implementation for the selected airlines. In T. Walker, A. S. Bergantino, N. S. Much, L. Loiacono (Eds), *Sustainable Aviation: Greening the Flight Path* (pp. 78–100). Palgrave Macmillan.
Reddy, T. L., & Thomson, R. J. (2015). *Environmental, Social and Economic Sustainability: Implications for Actuarial Science*. Institute of Actuaries of Australia.
Rizzi, F., Bartolozzi, I., Borghini, A., Frey, M. (2013). Environmental management of end-of-life products: Nine factors of sustainability in collaborative networks. *Business Strategy and the Environment*, *22*(8), 561–572.
Rotger, T. (2012). Aviation industry roadmap to sustainability. In D. Knorzer & J. Szodruch (Eds), *Innovation for Sustainable Aviation in a Global Environment* (pp. 116–121). IOS Press.
Schäfer, A.W., & Waitz I. A. (2014). Air transportation and the environment. *Transport Policy*, *34*, 1–4.
Sebastian, R. M., & Louis, J. (2021). Understanding waste management at airports: A study on current practices and challenges based on literature review. *Renewable and Sustainable Energy Reviews*, *147*, 111229.
Sheldon, P. J., & Park, S. (2011). An exploratory study of corporate social responsibility in the US travel industry. *Journal of Travel Research*, *50*(4), 392–407.
Skouloudis, A., Evangelinos, K., & Moraitis, S. (2012). Accountability and stakeholder engagement in the airport industry: An assessment of airports' CSR reports. *Journal of Air Transport Management*, *18*(1), 16–20.
Sledsens, T. (1998). *Sustainable Aviation: The Need for a European Environmental Aviation Charge*. European Federation for Transport and Environment (T&E), Brussels.
Stordal, F., Myhre, G., Stordal, E. J. G., Rossow, W. B., Lee, D. S., Arlander, D. W., & Svendby, T. (2005). Is there a trend in cirrus cloud cover due to aircraft traffic? *Atmospheric Chemistry and Physics*, *5*(8), 2155–2162.Sulej, A. M., Polkowska, & Ż, Namieśnik, J. (2012). Pollutants in airport runoff waters. *Critical Reviews in Environmental Science and Technology*, *42*(16), 1691–1734.
Syltevik, S., Karamperidis, S., Antony, J., & Taheri, B. (2018). Lean for airport services: A systematic literature review and agenda for future research. *International Journal of Quality and Reliability Management*, *35*(1), 34–49.
Upham, P., Thomas, C., Gillingwater, D., & Raper, D. (2003). Environmental capacity and airport operations: Current issues and future prospects. *Journal of Air Transport Management*, *9*(3), 145–151.
Upham, P. J., & Mills, J. N. (2005). Environmental and operational sustainability of airports: Core indicators & stakeholder communication. *Benchmarking: International Journal*, *12*(2), 166–179.
Wald, A., Fay, C. & Gleich, R. (2010). *Introduction to Aviation Management*. LIT Verlag.

Walker, S., & Cook, M. (2009). The contested concept of sustainable aviation. *Sustainable Development*, *17*(6), 378–390.

Weinhofer, G., & Busch, T. (2013). Corporate strategies for managing climate risks. *Business Strategy and the Environment*, *22*(2), 121–144.

Yang, X., & Rivers, C. (2009). Antecedents of CSR practices in MNCs' subsidiaries: A stakeholder and institutional perspective. *Journal of Business Ethics*, *86*, 155–169.

Yilmaz, N., & Atmanli, A. (2017). Sustainable alternative fuels in aviation. *Energy*, *140*, 1378–1386.

Yim, S. H. L., Lee, G. L., Lee, I. H., Allroggen, F., Ashok, A., Caiazzo, F., & Barrett, S. R. H. (2015). Global, regional and local health impacts of civil aviation emissions. *Environmental Research Letters*, *10*(3), 1–12.

Zaporozhets, O., & Blyukher, B. (2019). Risk methodology to access and control aircraft noise impact in vicinity of the airports. In T. H. Karakoç, C. O. Colpan, O. Altuntas & Y. Sohret (Eds), *Sustainable Aviation* (pp. 37–39). Springer.

10 Comparison of Airline and Railway Operations

CHAPTER OBJECTIVES

At the end of this chapter, you will be able to

- Understand the scheduling process in passenger railway operations.
- Understand the disruption management process in the rail industry.
- Know various railway schedule recovery aspects and approaches.
- Identify the similarities and differences in airline and railway operations.
- Understand the complexity and significance of airline and railway operations.

10.1 INTRODUCTION

In this chapter, sequential railway scheduling processes and railway schedule recovery solution approaches are presented. The schedule recovery problem for railways is addressed by considering timetable adjustment, rolling-stock rescheduling, and crew rescheduling aspects. The railway scheduling procedure is presented to discuss sequential and robust scheduling approaches prevalent in the rail industry. Moreover, disruption management aspects that are tackled in railways through timetable adjustments and rolling-stock rescheduling are also explored. Additionally, a comparison between the railways and aviation workflow is also discussed, with respect to the similarities in schedule construction and schedule recovery through different solution approaches. Also, the differences between railways and aviation along the lines of operation, technicality, recovery complexity, and issues relating to passengers are further delved into.

10.2 SCHEDULING IN RAILWAY INDUSTRY

Application of operations management tools and optimization models has been given considerable attention in various aspects of railway operations. In this chapter, the techniques that are used in passenger railway scheduling are presented with regards to scheduling and recovery aspects. In railway operations, among other reasons, disruption occurs when the daily train driver schedule becomes infeasible due to irregular operations on the railway network.

DOI: 10.1201/9780203731338-10

10.2.1 Sequential Scheduling

The traditional sequential planning process of resources in passenger railways, as in airline scheduling, first determines the network of stations to be served, followed by timetable construction, rolling stock planning, and crew scheduling. Railway resource planning can be classified into three stages: strategic, tactical, and operational (Huisman et al., 2005). In all these stages, different aspects of planning are considered. Sequential railway scheduling procedure is shown in Figure 10.1.

In the strategy stage, network-, line-, rolling stock- and crew capacity-planning is considered. These strategic plans are revised every 5–10 years for various reasons such as huge amount of capital is needed for infrastructure modifications and to address the probable variations in passenger demands. The tactical stage deals with the timetable construction aspects. At this stage, a new timetable is generated based on factors such as passenger demand and available infrastructure followed by the allocation of resources (rolling stock and train driver). Whereas, at the operational phase, timetable rescheduling and timetable recovery is considered. In timetable rescheduling, changes to the schedule, if needed, are carried out to address events such as construction work or repairs on a railway track. Timetable recovery is required if the day of operations is not proceeding as planned. For instance, if a rolling stock requires unplanned maintenance and is unavailable to operate or if a train is delayed

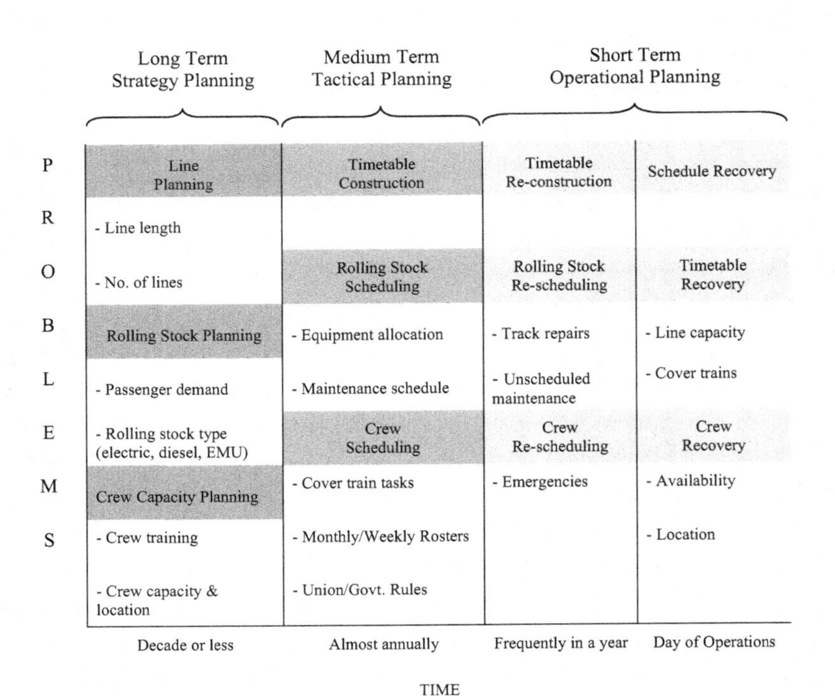

FIGURE 10.1 Sequential passenger railway scheduling procedure.

significantly leading to the breakdown of the timetable. The first two stages correspond to longer time horizons than the last stage.

The tactical stage deals with the timetable construction aspects. At this stage, a new timetable is generated based on factors such as passenger demand and available infrastructure followed by the allocation of resources (rolling stock and train driver). Whereas, at the operational phase, timetable re-scheduling and timetable recovery is considered. In timetable re-scheduling, changes to the schedule, if needed, are carried out to address events such as construction work or repairs on a railway track. Timetable recovery is required if the day of operations are not proceeding as planned. For instance, if a rolling stock requires unplanned maintenance and is unavailable to operate or if a train is delayed significantly leading to the breakdown of the timetable. The first two stages correspond to longer time horizons than the last stage. In all these stages, different aspects of planning are considered.

10.2.2 Robust Scheduling

The purpose of having robust schedules is to make the timetable and resources less sensitive to disruptions. Generally, robustness is achieved by incorporating buffer time during the construction of train timetables or rolling stock/crew scheduling. Buffer time is the surplus time given to complete a train task without making changes in the schedule. It keeps the schedule on time in case of low-impact disruptions. However, during severe disruptions buffer time alone is not sufficient to recover from schedule perturbations. In rail operations, robust scheduling has been given significant attention in different contexts such as robustness in timetable construction (Kroon et al., 2007) and rolling stock robustness (Huisman et al., 2005; Fioole et al., 2006).

10.3 DISRUPTION MANAGEMENT IN RAILWAYS

Similar to airline disruptions, a disruption in the railway schedule causes multiple problems for the operators. In railway operations, rolling stock unavailability and broken train driver pairings lead to the disruption of the train timetable. A train timetable, like an airline schedule, is constructed assuming disruption would not occur during the day of operations. However, disruptions occur frequently and lead to operational problems. The disruptions are solved sequentially starting from timetable adjustment, followed by rolling stock rescheduling and train driver rescheduling. Resources in railways, as in airlines, are interdependent on each other which causes disruptions to propagate through the network. This is called *knock-on* effect. Frequent causes of disruptions in the rail industry are infrastructure malfunctions, rolling-stock breakdowns, accidents, inclement weather, and so on. Disruptions with low level of impact sometimes can be handled by dispatching information such as overtaking trains, inserting an on-time train at an intermediate station, trains skipping stations (happens rarely), and through increasing residual capacity by canceling departures. Residual capacity is the difference between operational network capacity and the utilization of network at a given time. An illustration of a railway network with three lines (red, blue, and green) and 12 stations is presented in Figure 10.2.

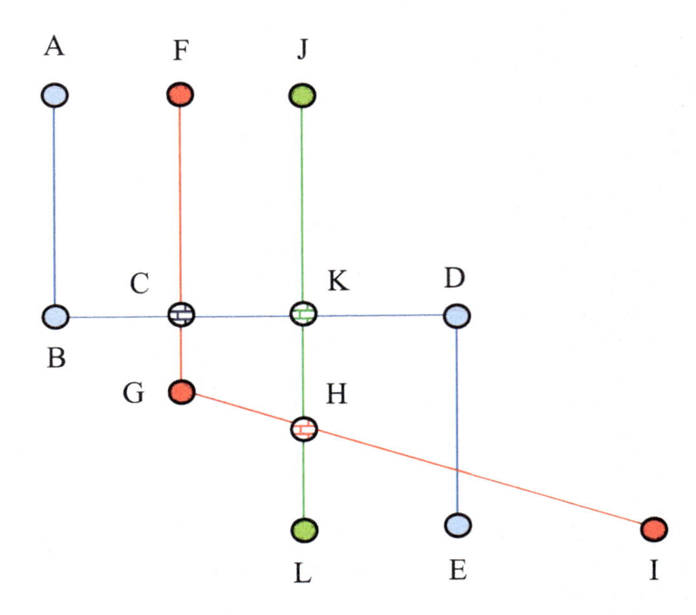

FIGURE 10.2 A railway network consisting of 12 stations and three lines (red, blue, and green).

A line is a route between stations which is operated with a specific frequency. On every line, there are stations represented by letters such as A, B, C, and so on. The blue line starts from station A followed by stations B, C, D, and E whereas the red line runs from station F to I through stations G and H. The green line operates from station J and before terminating at station L it passes through station K.

In the network shown in Figure 10.2, in some instances, the lines intersect. An intersection in the network is a station from which two or more lines run through. Red and blue lines intersect at station C, red and green lines intersect at station H and blue and green lines intersect each other at station K. These intersections serve as hubs in the network where passengers get an opportunity to change lines to reach their desired destination. Also, hubs give flexibility in the strategic planning phase and provide different recovery options at the operational phase.

For instance, based on Figure 10.2, in case of a disruption between stations C and G which is a segment (i.e., two consecutive stations) in the network, the network will be affected. Station C serves as a hub for red and blue lines while station G is on the red line. The segment C-G is used in routing the trains to connect stations F, H, and I on the red line, stations A, B, D, and E on the blue line and stations J, K, and L on the green line. A train route including stations I, H, G, C, and F (the red line) is published in the timetable and is allocated with the rolling stock and the crew to serve this route. A disruption due to inclement weather in segment C-G of this route makes this route non-operational for a significant time. Also, other routes which use this segment would be directly affected and become infeasible to operate on. The trains using segment C-G in their routes are required to be diverted through other segments which would lead to congestion in the remaining operational network. In this scenario, to run the operations as planned, schedule recovery is required by carrying out timetable adjustments, rolling stock rescheduling and crew re-pairing.

In railway schedule recovery operations, among other approaches, delay is managed through timetable changes without rescheduling the rolling stock and crews (Schöbel, 2007; Ginkel & Schöbel 2007; Heilporn et al., 2008). Disruption management considerations in the rail industry are presented in the following sections.

10.3.1 Timetable Adjustments

In case of disruptions, the schedule recovery process starts by making changes in the timetable. This is achieved by delaying, re-routing, or canceling trains. Timetable re-scheduling is carried out after assessing the impact of disruption in the schedule. Evaluation of the impact of disruption is based on factors such as:

- Number of affected passengers.
- Probable recovery time.
- Number of affected resources (rolling stock and train drivers).
- Affected segment(s) of the network.
- Time of the day.

After a disruption has occurred, during timetable rescheduling, train conflict resolution is an important issue in the recovery process (Törnquist & Persson, 2007). In the conflict resolution problem, the aim is to find a schedule where no two trains are on the same block section at the same time. A block section is a section of tracks between two signals in a network. Similarly, the approaches related to the allocation of track sections in a train network is also critical in ensuring the schedule recovery (Lusby et al., 2011).

To address recovery from the delays in rail operations, De Shutter and Boom (2002) apply a model predictive control (MPC) framework. The MPC has a prediction horizon, a receding-horizon procedure, and provides updates. The main aim of MCP is to optimally recover from past delays or known or expected future delays by breaking connections or by letting the trains run faster, where cost is associated with each control action. The approach is considered on a network of four stations that were connected by six single tracks for a 60-min timetable period. Two trains are considered for different routes on this network. Re-routing, adapting the time schedule, and speed control are not considered.

Törnquist and Persson (2007) address the problem of rescheduling traffic in an *n*-tracked network. The rescheduling problem is considered for multiple-tracked segments (two adjacent stations). The objectives of the problem are defined as goal functions. The first objective was to minimize the total delay of the traffic and the second objective minimizes the total cost associated with the delays. Four strategies are evaluated regarding optimality, speed, and problem size. Strategy one allows swaps of tracks by maintaining the order of trains within the segments while strategy two works the same as strategy one, but it ignores the sequence of trains in the segment. Strategies three and four allow swaps within specific segments. These strategies are applied in the Swedish railway traffic network. The network has 57 double-tracked and 125 single-tracked train segments including 169 stations and meet points. Train lengths are not considered in the problem and some track segments are considered bi-directional. The data used is the timetable of one day on this network for 92 freight trains and 466 passenger trains. Four types of trains are used: high-speed passenger trains, intercity trains, fast freight trains, and slow freight trains. A disturbance during the morning peak hour is randomly selected at a hub station. A hub is point where more than one line (routes through stations) cross. In total, 90 scenarios with different characteristics are generated and tested on four strategies considering time horizons of 30, 60, and 90 min.

10.3.2 Rolling Stock Re-scheduling

Prior to rolling stock rescheduling, it is assumed that the timetable is adjusted in the disrupted scenario. Maintenance issues are important as rolling stock needs to be in good condition all the time and hence maintenance checks are conducted after a certain number of kilometers or days (Maróti & Kroon, 2005; Maróti & Kroon, 2007). The recovery can also be achieved by finding a reinsertion plan of a train line by minimizing the latest time of insertion (Jespersen-Groth et al., 2009). The reinsertion problem occurs when all train departures of train lines are canceled for a specific period. Using this approach, specific trains are taken out from the schedule and the problem is solved computationally to recover the schedule.

The rolling stock problem can be addressed in different ways. For instance, Budai et al. (2010) consider the rolling stock as a balancing problem (RSBP). The RSBP occurs between the timetable and the rolling stock schedule in which rolling stock duties and the duties in the timetable do not remain as planned due to changing circumstances during operations. A duty is a sequence of activities performed by a rolling stock or train driver. The aim is to correct these off-balances by modifying the schedule so that it can be implemented in practice. The problem is addressed heuristically to identify several elementary balancing possibilities in the schedule followed by selecting some of the balancing possibilities (computed earlier) such that it leads to a new rolling stock schedule without off-balances (Budai et al., 2010). This is an iterative approach, in which either switching or re-routing is carried out to decrease the number of off-balances. Switching pairs rolling stock units of different types whereas the re-routing modifies the internal structure of the schedule to solve the off-balance problem.

Another approach is to consider the Rolling Stock Rescheduling Problem to update the current circulation of rolling stock to changes in the timetable (Nielsen et al., 2012). In this, rolling stock circulation is updated every time a timetable change is encountered due to disruptions. Uncertainties regarding duration of disruptions and problem size are dealt with using a rolling horizon framework. In this approach, only rolling stock decisions within a certain horizon of the time of rescheduling are considered. Also, penalizing the cancellation of trains, modifications of the shunting process, along with end-of-the day rolling stock off-balances can be considered for recovery (Fioole et al., 2006).

10.3.3 Crew Re-scheduling

The crew re-scheduling problem is solved after the timetable is adjusted and rolling stock is rescheduled. Crew rescheduling in railways is less complicated than airline crew rescheduling (Rezanova & Ryan, 2010). In crew rescheduling, a train driver can operate different rolling stock whereas a pilot has the license to fly only one or two aircraft types.

The train driver recovery problem (TDRP) occurs when the daily train driver schedule becomes infeasible due to irregular operations (Rezanova & Ryan, 2010). The problem was solved using the concept of disruptions neighborhood. In this concept, first an initial neighborhood is generated in which a small number (compared to the resources in the daily schedule) of disrupted resources and tasks are included. Every resource within the disruption neighborhood is assigned a recovery duty, which is a sequence of activities allocated to a resource as a substitute of its original duty. In initial neighborhood, if feasible solution is not found then the neighborhood is expanded. Expansion of the neighborhood is carried out either by adding more resources and/or extending the recovery period (time span between the recovery start time and end time). At this stage delaying of tasks is also considered. In this method, TDRP is aimed at finding a set of feasible duties for train drivers in the disruption neighborhood, such that tasks within the recovery period are covered and duties outside the recovery period remain unchanged. TDRP was formulated as a set partitioning problem with the objective to minimize the total cost of the recovery solution.

A crew rescheduling problem is addressed to a disruption caused by track maintenance (Huisman, 2007). In this scenario, the objective is to minimize the total costs of the crew duties. The approach considers replacing part of the original duties with different duties and for each original duty lookalike duties are generated. This approach was applied to medium and large cases of disruption on the largest passenger railway operator in the Netherlands (Huisman, 2007). In another instance, the operational crew rescheduling problem was addressed to minimize the cost associated with modification of duties and cancellation of a task (Potthoff et al., 2010). The core problem was considered as the subset of the original duties and tasks and few crew were included in the core problem to achieve recovery within a certain period. In this approach, neighborhood exploration was performed after some tasks and duties were left uncovered in the initial solution. As in airlines, integrated recovery approach is also considered in railways to recover the timetables and crew rosters simultaneously (Walker et al., 2005). In this approach, following a disruption in the schedule, simultaneous recovery of train and crew timetables in real time is considered by constructing a train timetable and crew roster to minimize the sum of the weighted train idle times along with penalizing the deviations from the existing schedule (Walker et al., 2005).

10.4 AIRLINE AND RAILWAY SCHEDULING: A COMPARISON

Airline scheduling and recovery procedures were presented in Chapters 6 and 7. In this chapter, railway timetable, train driver scheduling, and issues related with rail disruption are addressed. Therefore, in the following sections similarities and differences in airline and railway industries regarding scheduling and disruption management aspects are covered to get an overview of the operations of the two major global transport carriers. However, it is important to know that the comparisons between aviation and railway operations are generic and capture the industry practices. Region-specific aspects and regulations are not considered in these comparisons.

10.4.1 Similarities

The similarities in airline and railways are grouped in three categories: schedule construction, recovery procedure, and solution approaches.

10.4.1.1 Schedule Construction

Airline and railways generally follow sequential procedure of schedule and timetable construction, respectively. In this procedure, first, a network of airports (airline) or stations (railway) to be served is identified. It is followed by flight schedule construction in airlines, whereas in railways, a timetable is generated. Finally, allocation of aircraft and rolling stock routings, and generation of crew pairings (aviation) and train driver duties (railways) are generated. After the equipment type (aircraft in aviation and rolling stock in railways) is assigned, crews (captain and first officer in aviation and train driver in railways) are allocated duties in the schedule to operate the task.

10.4.1.2 Recovery Procedure

During disruptions, schedule recovery is also conducted in a sequential manner in airline and railways. The recovery process begins with aircraft or rolling stock recovery followed by re-pairing crew pairings and re-allocation of disrupted passengers. Due to stringent union rules and the regulations of international aviation bodies, crew recovery for both technical and cabin crews in airline operations is more complex than train driver recovery. Similarly, railway passenger recovery is much complex than airline passenger recovery because railway passenger destinations are unknown in advance. In the recovery procedure, implicit constraints such as aircraft or rolling stock maintenance and crew base restrictions are also considered both in airline and railway operations.

10.4.1.3 Solution Approaches

In airline and railway operations, both scheduling and recovery problems are primarily solved sequentially. Heuristic approaches and optimization models are used to solve scheduling and a range of recovery problems in airline and railway operations. The scope of recovery problems may vary from minor (less than 15 min) to major (more than an hour) delays. Regardless, the solution approaches used by the practitioners are based on various mathematical models and heuristics approaches. For fast and optimal solutions, the problems are solved computationally using optimization solvers such as CPLEX and Gurobi and solutions are obtained in seconds.

10.4.2 Differences

The differences in airline and rail operations are classified in four categories: operational, technical, recovery complexity, and passenger issues.

10.4.2.1 Operational

Scale of Operations

Size of operations in airline and railways vary significantly in many areas. Almost all major airlines (including subsidiaries) operate on domestic as well as international flights for e.g., Air India, Air New Zealand, Lufthansa, and British Airways. Whereas most passenger railway operators function regionally or nationally. However, some European rail operators such as Deutsche Bahn Fernverkehr (Germany), Statens Järnvägar (Sweden) and operators in North America, e.g., Amtrak (United States), Via Rail (Canada) also serve some destinations in their neighboring countries.

Infrastructure and Resources

Airlines operate on a massive scale as compared to railways and require considerable number of resources such as airports, aircraft, different crew types, and other logistics support functioning smoothly. Railways have different requirements such as train stations, rail tracks, rolling stock, crews etc. which are not as sophisticated and expensive as the resources required in the aviation industry. Due to industry regulations, infrastructural requirements in aviation operations are almost similar all over the world, whereas in railways it differs from country to country.

Reservation System

Airlines cannot operate profitably unless they know passenger demand, based on which an airline decides which flights to operate using an appropriate fleet type. An itinerary is generated at the time of reservation which includes description of travel date, day, and time. Railways, however, do not depend solely on the reservation criterion in deciding their operations. Rail operations do not necessarily require prior information of passengers traveling (except in some cases) whereas it is unlikely to carry out airline operations without such information.

10.4.2.2 Technical

Fleet Type

Airlines operate with different fleet of aircraft as compared to railways which do not employ a wide range of rolling stock or cabin cars. In rail operations, rolling stock types are limited as opposed to the range of aircraft fleets available in aviation operations.

State-of-the-Art Equipment

Technological advancements have made airline operations easy. Airlines require sophisticated state-of-the-art equipment and resources at different stages of their operations to function effectively and efficiently. Railways, however, mainly use sophisticated equipment for signaling purposes.

10.4.2.3 Recovery Complexity

Disruptions

As compared to railways, airlines have a significant number of disruption scenarios that lead to complex recovery. Whereas railway schedule recovery is complex in some respects (limited rail tracks) and easy (less expensive in cost) in others. In recovering a rail schedule, availability of tracks is a major issue which is not the case with airline schedule recovery. Due to this constraint, the order of trains is also considered during recovery. However, crew recovery is less binding in railways because a train driver can operate more than one type of rolling stock, unlike a pilot who is licensed to operate only one or two aircraft types which makes airline recovery difficult. Alternative transportation is easily available in case of railway disruptions unlikein airline disruptions.

Rules

During airline and rail schedule recovery, various rules such as equipment maintenance, crew operating hours, union, and government regulations make recovery difficult. Both, airline and railways, have different set of rules during disruptions.

10.4.2.4 Passenger Issues

Number of Passengers

Generally, airlines carry a greater number of passengers on a yearly basis than railways, except for countries such as China and India where rail connectivity is much better than air connectivity for commuting short distances.

Passenger Travelling Time

In most of the countries, passengers prefer to travel by air than by rail to save time because air travel is now affordable with deregulation of airlines in many countries.

A summary of comparison is given in Table 10.1. These comparisons on airline and railways are relative to each other and do not represent specific situations.

10.5 CONCLUSION

This chapter presents a comparison between airline and rail industry operations, exploring the points of similarity and variance between both industries. Aspects such as disruption management are given significant attention in the aviation industry compared to the disruption management in railways. And while schedule recovery has been focused upon in the railway industry of late, in aviation it has been a matter of import for decades. Some of the chief reasons for the significant delay in these timelines may be the limited competition among railway operators, governmental subsidies to the industry, and most importantly, low disruption costs associated with railways when compared to airlines. Other points of dissimilarity between the two industries include operational, technical, and other issues relating to passengers. However, there are also a number of similarities in aviation and railways, such as, schedule construction, type of recovery models, and solution approaches.

TABLE 10.1
A comparison of airline and railway operations

Parameter	Airline	Railways
Schedule Construction	Sequential	Sequential
Recovery Approach	Sequential	Sequential
Solution Procedures	Heuristics Optimization models	Heuristics Optimization models
Scale of Operations	Large	Medium
Operational Infrastructure	Mostly uniform	Varies
Resource Requirement	High	Moderate
Reservation System	Imminent	Optional
Equipment Fleet	Many	Few
Recovery Complexity	High	Varies
Rules	Many	Few
Travel time	Less	More

CHAPTER QUESTIONS

Q1. Discuss the objectives of strategic, tactical, and operational strategies in planning railway operations.
Q2. Identify two advantages and disadvantages in sequential and robust scheduling approaches in railway operations.
Q3. Based on Figure 10.2, in case of a disruption between stations H and L in the network, how will the network be affected? Identify three ripple effects of the disruption on the network.
Q4. For each parameter presented in Table 10.1, compare between aviation and railway operations regarding time and cost, as appropriate.
Q5. Discuss the similarities and differences between sequential airline scheduling and railway timetabling.

REFERENCES

Budai, G., Maróti, G., Dekker, R., Huisman, D., & Kroon, L. (2010). Re-scheduling in passenger railways: The rolling stock balancing problem. *Journal of Scheduling*, *13*(3), 281–297.

De Shutter, B., & van den Boom, T. J. J. (2002). Connection and speed control in railway systems – a model predictive control approach. Proceedings of the 6th International Workshop on Discrete Event Systems. Spain. 49–54.

Fioole, P-J., Kroon, L., Maróti, G., & Schrijver. A. (2006). A rolling stock circulation model for combining and splitting of passenger trains. *European Journal of Operations Research*, *174*(2), 1281–1297.

Ginkel, A., & Schöbel, A. (2007). To wait or not to wait? The bicrieteria delay management problem in public transportation. *Transportation Science*, *41*(4), 527–538.Heilporn, G., De Giovanni, L., & Labbe, M. (2008). Optimization models for the single delay management problem in public transportation. *European Journal of Operations Research*,

189(3), 762–774.Huisman, D. (2007). A column generation approach for the rail crew re-scheduling problem. *European Journal of Operations Research, 180*(1), 163–173.

Huisman, D., Kroon, L. G., Lentink, R. M., & Vromans, M. J. C. M. (2005). Operations research in passenger railway transportation. *Statistica Neerlandica, 59*(4), 467–497.

Jespersen-Groth, J., Potthoff, D., Clausen, J., Huisman, D., Kroon, L, Maróti, G., & Neilsen, M. N. (2009). In R. K. Ahuja, R. H. Möhring, & C. D. Zaroliagis (Eds), *Disruption Management in Passenger Railway Transportation* (pp.399–421). Springer.

Kroon, L., Dekker, R., & Vromans, M. J. C. M. (2007). In F. Geraets, L., Kroon, A., Schöbel, D., Wagner, & C. D., Zaroliagis, (Eds), *Cyclic railway timetabling: a stochastic optimization approach* (pp. 41–66). Springer-Verlag.

Lusby, R. M., Larsen, J., Ehrgott, M., & Ryan, D. (2011). Railway track allocation: Models and methods. *OR Spectrum, 33*, 843–883.

Maróti, G., & Kroon, L. G. (2005). Maintenance routing for train units: The transition model. *Transportation Science, 39*(4), 518–525.

Maróti, G., & Kroon, L. G. (2007). Maintenance routing for train units: The interchange model. *Computers and Operations Research, 34*(4), 1121–1140.

Nielsen, L. K., Kroon, L., & Maróti, G. (2012). A rolling horizon approach for disruption management of railway rolling stock. *European Journal of Operational Research, 220*(2), 496–509.

Potthoff, D., Huisman, D., & Desaulniers, G. (2010). Column generation with dynamic duty selection for railway crew scheduling. *Transportation Science, 44*(4), 493–505.

Rezanova, N. J., & Ryan, D. M. (2010). The train driver recovery problem – A set partitioning based model and solution method. *Computers and Operations Research, 37*(5), 845–856.

Schöbel, A. (2007). In F. Geraets, L., Kroon, A., Schöbel, D., Wagner, & C. D., Zaroliagis, (Eds), *Integer Programming Approaches for Solving the Delay Management Problem* (pp.145–170). Springer-Verlag.

Törnquist, J., & Persson, J. A., (2007). N-traked railway traffic re-scheduling during disturbances. *Transportation Research Part B, 41*, 342–362.

Walker, C. G., Snowdon, J. N., & Ryan, D. M. (2005). Simultaneous disruption recovery of a train timetable and crew roster in real time. *Computers and Operations Research, 32*, 2077–2094.

Appendices

APPENDIX A: EASA MEMBER STATES

1. Austria
2. Belgium
3. Bulgaria
4. Croatia
5. Cyprus
6. Czech Republic
7. Denmark
8. Estonia
9. Finland
10. France
11. Germany
12. Greece
13. Hungary
14. Iceland
15. Ireland
16. Italy
17. Latvia
18. Liechtenstein
19. Lithuania
20. Luxembourg
21. Malta
22. Netherlands
23. Norway
24. Poland
25. Portugal
26. Romania
27. Slovakia
28. Slovenia
29. Spain
30. Sweden
31. Switzerland

APPENDIX B: NETWORK CONCEPTS

A network is a pair of sets (N, A), where N is a set of nodes, and A is a set of arcs. An *arc* joins two nodes in the direction from i to j i.e., the arcs connect nodes together to create a network. Nodes i, j are said to be adjacent if they are connected through an arc. When i and j are represented where i always precedes j in time then it is considered as an ordered pair (i, j) with i and j representing *tail* and *head* of an arc, respectively. Whereas, i and j are joined by an *edge* if they are unordered i.e., nodes i and j can be used either from i to j or from j to i. A network is called a *directed network* if it contains only arcs, an *undirected network* if it has only edges, and a *mixed network* if it consists of both arcs and edges.

In a network there can be more than one arc (in the same direction) or edges between adjacent nodes. Such arcs or edges are called *parallel arcs* or *parallel edges*. In a network with parallel arcs/edges, each arc/edge is given a unique arc/edge identifier in the network. For instance, in Figure A.1, nodes A and B are connected by three parallel arcs each representing a unique resource. A red arc (i_1, j_1) represents an aircraft whereas, a captain (i_2, j_2) and a first officer (i_3, j_3) are represented by blue and green arcs, respectively. Similarly, all other parallel arcs in the network are represented by unique identifiers. An undirected network which has no self-loops is called a *graph* (if it has no parallel edges) or a *multigraph* otherwise. A self-loop is an arc/edge which joins a node with itself. The *degree* of a node in a network is the number of arcs/edges incident at it. In a directed network, the *indegree* of a node is the number of arcs incident into it and the *outdegree* of a node is the number of arcs incident out of it.

Based on the problems instances presented in the book, a directed network consisting of five nodes and three types of arcs is shown in Figure A.1. In airline scheduling and disruption management scenarios, a node represents a task, and an arc represents a resource. The disruption neighborhood network is constructed in such a way that nodes which are connected through arcs represent a *sequence* of tasks or activities to be performed by a resource. Each of these sequences is called a *path* and is represented as a *column* in the problem formulation.

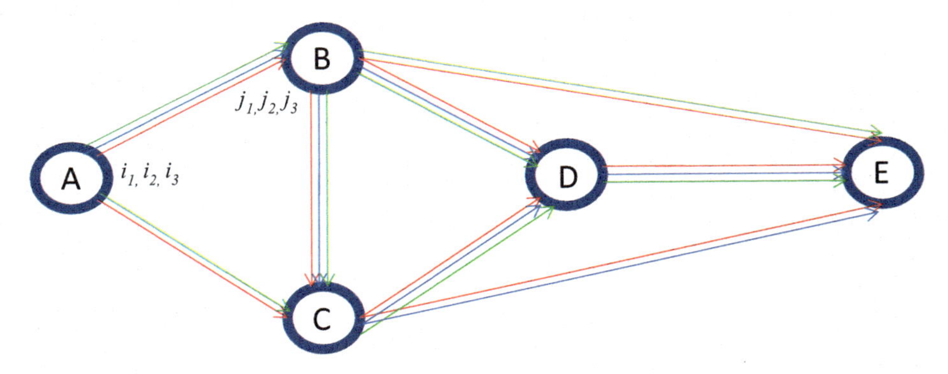

FIGURE A.1 A network of nodes representing airline resources.

In column generation technique, all the possible paths or sequence of duties, which can be performed by a resource between start node and end node, are generated. For instance, in Figure A.1 for an aircraft (red arcs), between nodes A (*start node*) and E (*end node*) there exists six paths as presented below.

- Path 1: *A-B-E*
- Path 2: *A-B-D-E*
- Path 3: *A-B-C-D-E*
- Path 4: *A-B-C-E*
- Path 5: *A-C-D-E*
- Path 6: *A-C-E*

Each of these paths represents a column in the problem for an aircraft if the port entry and exit conditions are not violated. Similarly, in Figure A.1, between a start node and end node there are three paths for a captain (blue arcs).

- Path 1: *A-B-D-E*
- Path 2: *A-B-C-D-E*
- Path 3: *A-B-C-E*

Each of these paths represents a column in the problem for a captain if crew port entry and exit conditions are satisfied. Similarly, in Figure A.1, between a start node and end node there are four paths for the first officer (green arcs) subject to meeting the port entry and exit criteria.

- Path 1: *A-B-E*
- Path 2: *A-B-D-E*
- Path 3: *A-B-C-D-E*
- Path 4: *A-C-D-E*

These paths can be generated either through *depth-first-search* (DFS) or *breath-first-search* (BFS) algorithms. DFS technique starts from the root node (start node) of the network and explores as far as possible (until the end node) along each branch before going back to its predecessor node. However, in BFS, after root node all the neighboring nodes are explored first then for each of these nodes their unexplored neighbors are explored until it reaches the end node.

APPENDIX C: AIRCRAFT CLASSIFICATION

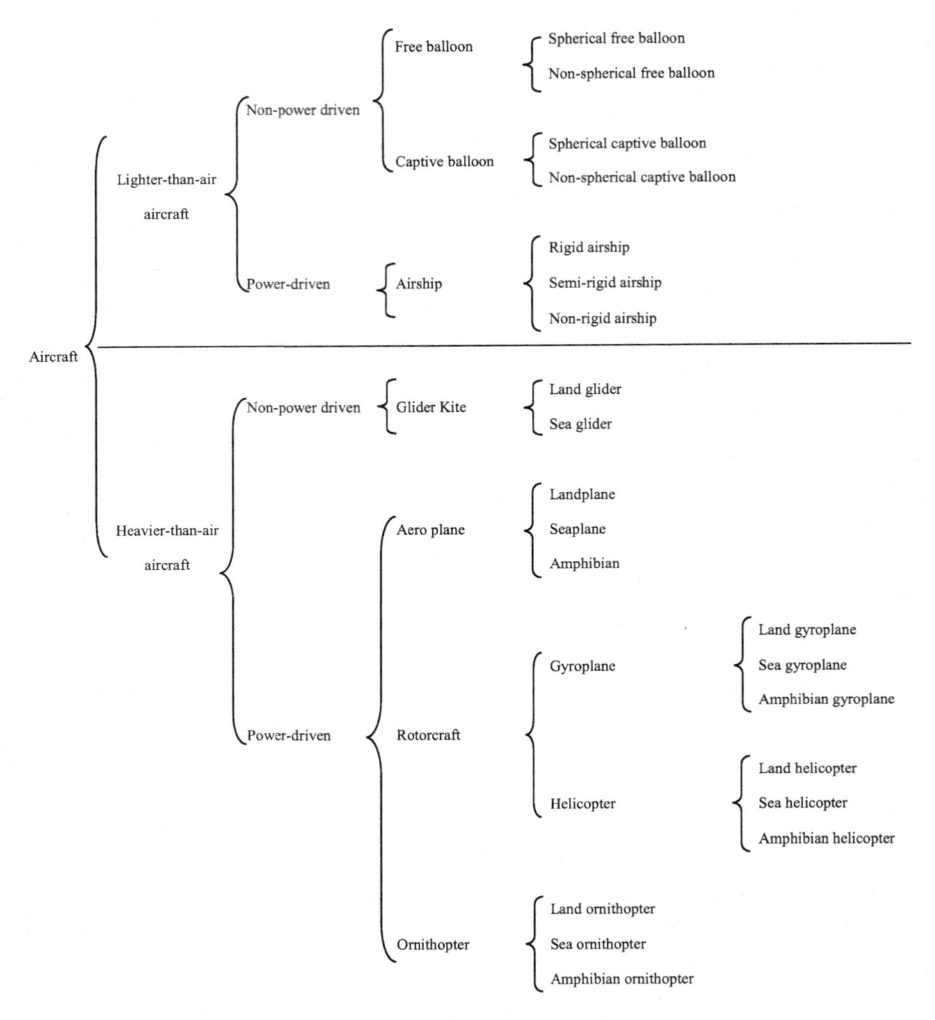

FIGURE A.2 Classification of aircraft types.

APPENDIX D: BASIC AVIATION TERMINOLOGY

- *Bid lines*: Sequences of pairings assigned to crews. Bid lines are subject to union and/or industry legalities.
- *Block time:* The time from aircraft engine start up to shut down.
- *Broken pairing*: Crew pairing that is affected by the disrupted schedule but have not yet been repaired.
- *Crew base*: Designated stations where crews are stationed (often a hub). A *crew base* is a city where crew pairings begin and end.
- *Crew pairing*: Sequence of duties, starting and ending at the same crew base that satisfies all contractual restrictions (also called *legalities*). A *pairing* usually consists of several duty periods and is preceded and followed by a rest period. A pairing must start and end at the same airport.
- *Deadhead*: A situation when crew members travel as passengers on a flight leg to position themselves for their next flight or to head home at the end of a duty period.
- *Duty*: Working day of a crew. It consists of a sequence of flights, where the arrival station of a flight is the departure station of the next flight. A duty is subject to regulatory (such as FAA) and company rules.
- *Duty period*: Period between reporting for a task and being released from that task. It is preceded and followed by a rest period. A duty period can be viewed as one workday.
- *Elapsed time*: Time that elapse between the beginning of a duty and the end of the duty.
- *Flight duration:* Includes the block time and the turnover time on the ground.
- *Flight leg:* A nonstop flight. Also, known as a flight segment.
- *Hub*: A port with high activity that connects flights from other ports.
- *Move-up crews:* The crews that can potentially be swapped in the operations.
- *MinSit* and *MaxSit*: Minimum and maximum connection time between two consecutive flights for a crew.
- *Minimum ground turnaround time:* The minimum time an aircraft must spend at an airport between any two flights.
- *Reserve crew*: Crew on call, staying at home ready to work, if required. A reserve crew has minimum guaranteed paid hours even if no duty is performed.
- *Sit connection*: A connection within a duty.
- *Switch delay*: A delay due to an aircraft change.
- *Sit time*: The time between two consecutive legs within a duty.
- *Uncovered flights*: Set of flights without assigned aircraft and/or crew.

Index

A

aerodrome standards 36, 38
airbus 62
aircraft certification 34–5
aircraft engineering 34–8
aircraft noise 77, 153, 155–6, 158, 160
aircraft recovery 148
aircraft routing 84, 86, 88, 97–8, 106, 135
airline operators 5–6, 16, 19, 26, 28–9, 33, 37, 45–6, 48, 66, 75, 78, 163, 166
airline schedule recovery 102, 104, 123–4, 142–3, 145–6, 148, 182
airline scheduling 1, 83–4, 89, 97–8, 111, 174, 179
airline supply chain 51, 61–2, 64, 68
air navigation services 34–5
air pollution 154, 158
airport capacity 75–7, 161, 131
airport categories 9
airport codes 9, 11, 15, 40–4, 87
airport location 15, 18–19, 21, 23, 64, 77
airport service quality 78
airport supply chain 66–8
air safety 36–8
airside operations 72–3
air traffic control 4, 11, 14, 44, 66, 68, 72, 102
airworthiness 14, 34–6
arrival port 29, 72, 87, 102–3, 129, 131–3, 147
aviation industry trends 76
aviation operations 1–8, 10, 14, 28–9, 33–7, 46, 51, 61, 66–7, 71–2, 75–6, 153–60, 162–6, 181
aviation policy 45
aviation supply chains 61, 65–6, 68, 159
aviation sustainability 153

B

baggage handling 14, 27, 29, 66, 68, 72, 75
baggage reclaim 75
Brundtland Commission 153
busiest airports 15–16, 19, 21, 23, 24, 44

C

catering services 61–3, 65, 68
civil aviation industry 7, 25, 33, 36, 61, 156
climate change 155, 158
column generation 111, 115, 133, 142, 187
connection networks 105, 110
crew pairing 84, 86–8, 92–6, 98, 110, 112–16, 137, 140, 143, 148, 180, 188–9
crew recovery 105, 114–18, 148, 174, 180, 182
crew rostering 83–4, 88, 114–15, 117
crew scheduling 61–2, 84, 86, 88, 97, 107, 111–12, 114, 117, 135, 174–5

D

departure delays 114, 117
departure port 29, 72, 87, 102–3, 127, 133
Directorate General of Civil Aviation 34
disrupted aircraft 105, 128
disrupted crew 105
disruption management 61–2, 101, 105–6, 114, 123, 148, 173, 177, 180, 182, 186
disruption neighborhood 103, 123–8, 135, 137, 140–3, 145, 179, 186
distribution network 53, 60
domestic airports 38-9

E

Eastern Hemisphere 12–13
emissions 7, 153–7, 159–60, 162, 164–5
entry condition 125, 137, 139–40
entry port 125, 133, 134
entry time 125, 127, 133–5
environmental sustainability 155, 157, 159, 161–4
equipment manufacturers 61–2, 68, 159
European Aviation Space Agency 14
exit condition 125, 137, 139
exit port 125, 128, 133–4, 140
exit time 125, 128, 133–4, 140
extended neighborhood 150

F

Federal Aviation Administration 14, 78, 86
first officer pairings 92, 140
fixed flight schedule 114
fleet assignment 84–6, 97–8, 112
fleet size 25, 45, 47–8
flight cancellations104, 106, 114, 116
flight schedule 61, 84–5, 104, 109–11, 113–15, 117, 119, 180
freight operators 27, 65–6, 72
fuel supplies 62, 65
full-service carriers 25, 46, 77

G

gate allocation 27, 66–8, 73, 76–7
global aviation operations 5, 76, 162

H

heavier-than-air flight 2, 5
heuristic approaches 120, 180
hub-and-spoke network 117

I

impact of Covid-19 28
initial neighborhood 125–7, 129–32, 145, 149, 179
International Air Transport Association 33
international airports 21, 23, 38, 40, 45
International Civil Aviation Organization 35, 155–6, 158, 162

L

landside operations 72, 79
lighter-than-air flight 2
low-cost carriers 21, 25, 73

M

maintenance checks 86, 104, 178
Ministry of Civil Aviation 34

N

National Aviation Space Agency 6
network creation 141
network flow 105–8, 114, 117–18

P

passenger recovery 104–5, 113, 118–19, 180
passenger volume 6, 24, 28, 75, 79, 157, 160
point-to-point network 25
powered airplane 2
procurement process 54–6
pull system 59, 63
push system 59, 63

R

railway disruption management 123
railway operations 173, 175, 180
railway scheduling 173–4
recovery neighborhood 140, 143, 147
recovery period 105, 108–9, 112, 115, 124–7, 130, 136, 143, 150, 179
regulations: airport 14; aviation 34
regulatory bodies 10, 35, 37, 61–2, 65–8, 78, 162
rolling stock rescheduling 173, 175, 177–8
rolling time horizon 143
runway operations 71

S

social sustainability 159, 166
supply chain management 54
sustainability reporting 162
sustainable operations 6, 56, 58, 153–4, 161–2, 164–5

T

time-band networks 105, 109
time-space networks 105–6
timetable adjustments 173, 177
traffic conference areas 11, 40
train driver recovery 123, 179–80
training and licensing 44
triple bottom line 153

W

waste management 160, 165
water pollution 158, 160
Western Hemisphere 12–13